MINDFUL SCEPTICS
GUIDE TO...

saving the koala

Mark Dangerfield
Christopher M Scott

Mindful Sceptics

Published by Mindful Sceptics

Springwood, NSW 2777, Australia

mindfulsceptics.info

AI Use Statement

This book was conceived, written and edited by a human who occasionally used AI tools to assist with grammar, editing, summaries and focused searches.

Contents

Preface to the Series

Ever feel like you're drowning in information and controversy, unsure what to believe? You're not alone. In this age of information overload, finding clarity can seem like searching for a needle in a digital haystack.

But what if I told you that the key to navigating this chaos lies within you?

Welcome to the Mindful Sceptic Guides, your compass in the turbulent waters of modern discourse. These aren't your average self-help books or dry academic tomes. They're intellectual adventures designed to awaken your innate curiosity and sharpen your critical thinking skills.

Imagine being able to cut through the noise, to see past the spin and the hype. Picture yourself in conversations where your insights turn heads, where your words carry the weight of well-reasoned thought. That's the power of being a mindful sceptic, a power that's within your reach.

In the pages of *The Mindful Sceptic Guides*, you'll find a delightfully eclectic mix of crucial facts, thought-provoking ideas, and paradigm-shifting perspectives. We'll tackle big issues with a unique blend of scientific rigour and philosophical inquiry. From conservation to social justice, from food security to healthy diet, economic theory to personal growth, no topic is off-limits.

But here's the kicker. This isn't just about personal enlightenment. The stakes are higher than that. In a world grappling with complex, existential challenges, humanity's survival may depend on cultivating more mindful sceptics—people who can question dogma, seek evidence, and imagine new solutions to age-old problems.

Could you be one of these crucial thinkers? Could your ideas be the spark that ignites positive change?

There's only one way to find out. Dive into these guides with an open mind and a willingness to challenge your assumptions. Prepare for moments of revelation, ideas that will keep you up at night, and insights that will transform how you see the world and your place in it.

Are you ready to embark on this intellectual journey? To become a beacon of reason in a world that desperately needs it? Then, turn the page. Your adventure in mindful scepticism begins now.

Welcome to the Mindful Sceptic Guides. Let's change the world, one critical thought at a time.

Preface to this Guide

Imagine you're strolling through a shopping mall when you're approached by a smiling volunteer holding a donation bucket. "Save the koala!" she exclaims. Your heart melts a little. Who doesn't want to save these adorable, fuzzy creatures?

But wait. What if I told you that "saving the koala" is more than just a plea for conservation? What if it's a clever ruse, an emotional sleight-of-hand that obscures a far more complex reality?

Welcome to *The Mindful Sceptic Guide to Saving the Koala*, where we use Australia's iconic tree-hugger as a launching pad to explore some of humanity's most pressing challenges. We're about to embark on a journey that will challenge your assumptions, tickle your intellect, and maybe even change how you think about conservation.

Saving the koala, the panda, the polar bear, or any other creatures with fur creates an endorphin-enhancing response in humans because we like these creatures a lot. They are cute, and that suckers us into many an irrational response. In these pages, we'll unpack the 'cuteness conundrum' and explore how this emotional response shapes, and sometimes misshapes, our approach to environmental issues. And we'll dive deep into the thorny decisions humanity faces in balancing conservation with other crucial needs.

But don't worry. This isn't a doom-and-gloom manifesto. Our goal is to be myth-busting and inspirational in equal measure. We'll challenge conventional wisdom, yes, but we'll also uncover new ways of thinking about our relationship with nature.

So, whether you're a die-hard conservationist, a curious sceptic, or somewhere in between, I invite you to join me on this intellectual adventure. Prepare to be surprised, challenged, and inspired. By the time we're done, you'll never look at a koala or a conservation campaign in quite the same way again.

Ready to become a mindful sceptic? Let's dive in; the koalas are waiting, and so is a new way of thinking about our world.

Chapter One

Why we are besotted with cute animals

Gogglebox is a popular TV show. The Australian version, currently in its 18th season, hits half a million viewers per episode. It turns out that 'watching people on TV watching TV' is a clever and cheap way to capture the intensity of human emotions that the producers of television entertainment have wet dreams about. For example, we watched people watching a show about a zoo in the UK where a giraffe, pregnant for 14 months, gave birth on film. The 60-kilo youngster entered the world from a great height and lay still on the ground for an eternity. Was it alive? Did the fall break its neck? Was it stillborn? Oh my god, will it breathe? Breathe, please breathe.

Look! There, there, its nostrils are twitching. It's alive. Thank heaven it lives. Pass the tissues.

I kid you not. These were the reactions of the Goggleboxers and, no doubt, many thousands of viewers who saw the show when the emotion-churning scene went to air in its original program and on Gogglebox. It's why both sets of producers made and chose it. They knew there would be a profound reaction, and they knew they had gold as soon as the birthing footage was in. Rather than scurry along with the psychology of the entertainment industry, what does a mindful sceptic make of such reactions? What questions do we ask? Let's start with what makes humans so attached to animals and birth.

Consider the giraffe. Back in the day, we ate them. At least we would if we could catch them. An adult giraffe is hard to pull down with primitive tools, and the more manageable youngster has its mother with dinner plate-sized hooves on the end of long levers to help protect it. Even lions find it hard to kill a giraffe, so no doubt our ancestors did, too, but they would have tried. Cuteness did not overcome the desire to eat.

Somewhere along the way, charm grew or, as seems more likely, hunger withdrew. As soon as we had agriculture and food supply chains, adding fury creatures with wide eyes into our lives was much easier. We even started to keep pets, perhaps in response to our insatiable need for human relationships, especially between parent and child. How many lap dogs are surrogate kids that behave themselves? Cuteness in animals is because they can look like babies. The cutest pets tend to be small, have forward-facing eyes, are unconditionally needy, and fluffy or, as in the case of the giraffe on Gogglebox, newborn. Many of these attributes can generate more satisfaction than the real thing and are excellent surrogates for people who have grown-up children or none at all. This is the key. Animals are just like babies.

We find certain land animals attractive. And it sticks thanks to that unconditionality that all animals have, even the giraffe caged in a zoo. The animal is controlled and cannot hurt us. As visitors, we can smile, sigh, say squishy words, whiff the dung and move on. In return, it looks at us with those doughy eyes, and we think it loves us. It doesn't. Animals in zoos are caged; they don't even care that much for their keepers, even when they bring a regular supply of food.

We love animals, the furry ones, not the yukky creepy crawly ones. And indeed, this reality is of little consequence. All it means is that we will have pets, keep animals in zoos and watch David Attenborough documentaries. Not many of us can go past a dog's unconditional affection, even a cat's pretend aloofness. And then they are soft and cuddly, wow. This love is hard-wired. It is not going away.

Many viewers on or off Gogglebox would starve before they could knock the baby giraffe on the head and roast its leg. And if they were mad enough to do it, they would go to prison and suffer a slow death on social media. So we love them, the creatures with the wide eyes. It is a given. Now, let's think about what that means. Here are a few of the consequences:

1. There will be many pets, just shy of 90 million dogs in the US alone and growing by over 1 million annually.

2. We prefer to protect koalas because you can hold them, and their fur is soft, but maybe not polar bears so much, and elephants less than bears.

3. We will unlikely ever consider it essential to protect insects even though they are rapidly declining in numbers and species.

4. We will have to ignore that before we ate its rump, the cow was cute(ish), and that venison was... no, you can't say it.

Even if you go past the apparent contradictions, we have a problem. Our perception is off.

Chapter Two

Misleading claims for the future of koalas

Sometimes, people make impossible claims that make it hard to be objective, especially when the claim is designed to tug at emotions.

Telling lies for effect

A television advert aired in 2023 from Australian insurance company AIG under their NRMA brand made this claim:

> *To ensure the future of koalas, we are planting a tree for every new home insurance policy.*

What a complete load of cobblers. The declaration 'to ensure' is a complete and total lie. There is no way planting trees at this scale will make any difference to the survival of this species. This would be true even if the koala were in real danger of extinction—it isn't, by the way.

Let's run some numbers to uncover the lie. There are roughly 9 million private dwellings in Australia. Let's be generous and assume they all need insurance and have owners prepared to pay the premiums. We can also be generous to NRMA, one of many insurers in Australia, and say the brand has 20% of the home insurance market. Then let's say that 10% of the market turns over each year with the opportunity for 'new' policies and NRMA grows their book at 5% each year—they would be stoked with that level of new business growth given most people who can afford it are already well insured.

The back-of-envelope sums give us nine million policies and 1.8 million for NRMA, booking 90,000 new policies yearly. So, what will 90,000 trees look like for a koala? It sounds like a lot. Ecological evidence from wildlife research tells us that roughly 20 mature eucalypt trees are needed by one koala. This is an arbitrary amount that combines what space they need, what shelter they must have from the elements and predators, and, of course, food supply. If the trees promised are planted and quickly mature, NRMA could supply trees for 4,500 koalas. This must be good news and much better than doing nothing. But is 4,000 koalas a lot?

We don't know how many koalas live wild in Australia. The scientific estimates vary widely, with the lowest at 43,000 and the commonly repeated estimate of 100,000 to a more generous estimate of over a million. So again, let's be kind to the NRMA marketers and assume the lower number, 43,000. Once the trees mature in 20 to 30 years, NRMA tree planting will provide enough food and shelter for 10% of the koala population. Again, not bad. They may save the species after all. Perhaps I was too hasty. Adding 10% to the population—although it could be as little as 4% or even 0.5%—may happen as long as we can also assume that the trees planted are the suitable species, that they are in habitats that koalas like, and that they grow fast.

We will also assume considerable skill in ecological restoration and that the people planting the trees have these skills so that suitable trees are planted in the right places with a high chance of survival. The NRMA pays the planting costs, and the koala has the patience to wait for the trees to mature. A 10% risk buffer in 20 years is noble. It's better than not trying, but the greenwash is palpable, for with or without those insurance trees, the koala will not go extinct any time soon. It's a generalist survivor, after all. More on this later. But there is more context to the NRMA claim.

In the summer of 2019-20, one of multiple wildfires to the north-west of Sydney incinerated 846,000 ha of wilderness forest. At a conservative rate of 200 trees per hectare, this fire damaged at least 169 million trees. Most of the burnt trees recovered through epicormic growth, the remarkable response of sprouting leaves on shoots from the tree trunk that many species of eucalyptus achieve when the canopy leaves are burnt, and there will be recruitment from seeds when the rain comes. It's good for future koalas but could be better for those alive now, assuming they could lumber away from the firefronts.

The 90,000 trees planted by NRMA, which sounds like such a huge number, is just 0.05% of the trees damaged in just one of the fires in NSW in 2019. Sorry to have to tell you, but '*to ensure the future of koalas*' is pure greenwash or hogwash, if you prefer. It is false, and so out of touch with the scale of environmental issues, it is not even funny. Get a grip, everyone.

The real problem with koalas

I don't like koalas. More strictly, I don't like people's responses to them from the '*Ah, they are so cute*' to the '*OMG, they are about to go extinct*'. In my view, they are neither cute nor about to shuffle off into oblivion.

The main problem for a mindful sceptic is that my response and the spin from the Australian insurer are normative. They are emotional, which in the hierarchy of objectivity is a step down from opinion and a long way short of evidence.

No matter. We should expect people to get their hearts involved in things; it makes the world go around, so I am told. Remember that cuteness is an innate trigger, and no amount of pragmatism on my part will counter that emotion. More difficult to handle is the lack of objectivity. Even a shallow look at the available evidence tells us that the koala will not go extinct soon, and certainly not in the next five minutes.

The fossil evidence suggests that after koalas (Phascolarctidae) divergence from terrestrial wombats (Vombatidae) 30–40 million years ago, as many as 15–20 species recognised as koalas have existed. The modern koala, *Phascolarctos cinereus*, which first appeared in the fossil record ~350,000 years ago, a little before *Homo sapiens*, is the only extant species of the Phascolarctidae. Ecological research tells us that *Phascolarctos cinereus* is a specialised feeder, prone to certain diseases and has been squeezed by genetic bottlenecks, especially from small founder populations in the southern parts of Australia.

Researchers also point out that the current estimates put the number of koalas in Australia at '*only 329,000 (range 144,000–605,000)*', and a continuing decline is predicted. Again, '*only*' is a classic normative word, an opinion. As for the evidence, an error range of plus or minus 300,000 is too coarse to make any disaster claims valid. The first task must be to tighten the estimates to something closer to the actual numbers and the absolute rates of change. We do not have enough evidence to make pronouncements on the current population sizes, distribution, and relative changes in koala numbers. There are good reasons for this lack of knowledge. The koala is hard to count. Individuals are cryptic and mobile, often moving from one tree to the next and from one woodland patch to another. Wildlife ecologists need help to count the herds of antelope and elephants on the open grasslands. Koalas are small, isolated dots in the canopy of forest trees. Reliable counting is difficult.

As I was mulling over this puzzle for the umpteenth time, I heard a chat on the radio where some journalists were commenting on the devastating consequences of COVID-19 for the $60 billion Australian tourism industry. The gist of the conversation was that Australians are unlikely to take up the slack created by the loss of the Chinese market by tourism from the locals. They thought Australians were just not excited by the wildlife they grew up with, unlike the overseas tourists who are fascinated, often enough to travel thousands of kilometres to see them.

Now, this is interesting. It suggests to the cynical sceptic that all the koala brouhaha from state and Federal governments here in Australia over koala conservation has little to do with extinction. It is about attracting foreign tourists back to a market designed for them rather than for the locals. That $60 billion represents a lot of jobs, including in regional areas. It is money that matters. Evidence of extinction, not so much.

Objectivity

I'm sure you would say that you are impartial and *not influenced by personal feelings or opinions in considering and representing facts*. Indeed, you would say that objectivity is a feature of your everydayness. At least, this is what our parents and teachers taught us to say. However, these good folks often forget to mention that being objective takes effort and a lot of it. To be objective, a person must know or have access to kernels of truth, irrefutable information, that is, known or proved to be true. In short, objectivity requires evidence.

This means that becoming and remaining an objective person takes constant effort to assemble and evaluate evidence. It means knowing a fact when you see one and where to look for those you do not have to hand. It takes rigour.

Evidence should be everywhere in this world of internet searches and virtual assistants turbocharged with AI bots. However, just because an important or famous person says something or information is reported on commercial media does not make it a fact.

I am sure you have seen on the news a windswept reporter wearing a logo embossed gore tex jacket standing in front of a collapsed shed. There had been a supercell storm. An event so powerful that '*the town was all but destroyed*'. Only wait a minute. The '*town*' was home to 20 people; it was a hamlet at best where most of the houses were still standing. Pan out from the close crop of the flimsy shed, and the destruction was hard to see. There is learning, skill and experience in ensuring that information is factual, reliable and accurate, not to mention relevant. Access to information is only the first part of objectivity. There is information in profusion, and some of it is irrefutable. The shed was in a mess. The problem is there was no town and no information beyond the close-up of the shed to indicate destruction. We have a visual fact and a host of loose interpretations and opinions. Still, we trust the reputation of the news channel and believe that the reporter was indeed in central Queensland. But she could have been in a studio with a wind machine and a green screen. One irrefutable fact—an image of an upside-down shed—is insufficient evidence.

Objectivity requires numerous validated facts that must override a host of deep, ingrained and often long-standing feelings that are part nature, part nurture and part sheer bloody-mindedness. Objectivity means not just accumulation but the ability to sift through facts and put them into context, an evaluation process.

Now, we are lurching rapidly toward some serious effort. Not only must we ask Alexa or say '*Hey Google*' for the relevant facts, but we cannot take what the virtual assistants on the smartphone say as gospel, not even ChatGPT. Another step to objectivity requires us to evaluate the facts for their efficacy—are they true—and what that piece of information brings to the challenge or puzzle we are trying to be objective about.

Knowledge is about more than just committing facts to memory. It requires an additional step: relevance.

Back to the extinction example

The rhetoric is that the koala is in danger of going extinct and, therefore, we must do all we can to protect it. Australians understand what this means, but it doesn't have to be the koala; it could be the lion, tiger, African elephant or the ghost orchid, *Dendrophylax lindenii*, known only from a handful of tiny sites in Oxfordshire, England. Feel free to insert whatever species you are told is about to expire into the extinction clause. The challenge is being objective when presented with such a proclamation of irreversible loss.

Begin with getting yourself across the suitable, irrefutable and relevant facts. And here is the initial challenge. What facts are these? Given the edict, it would have to be facts about how many koalas are alive, what has happened to their numbers in historical and recent times, and evidence for the presence and veracity of threats to their continued existence. In other words, you need to be aware of and able to evaluate the population dynamics of koalas and the key drivers of those dynamics before you can be objective about the clause regarding extinction.

Curiously, the Australian government factsheet on *Koala populations in Queensland, New South Wales and the Australian Capital Territory and National Environment Law* does not say how many koalas there are in Australia. It does say koalas live wild across a range that stretches the entire eastern seaboard of the continent. Similarly, in New South Wales, the Australian state home to 7.5 million people, the NSW government's $44.7 million *NSW Koala Strategy* does not mention how many koalas there are in NSW. Earlier, I quoted research estimates that put the number of koalas in Australia at 329,000, ranging from 144,000 to 605,000. The official documents fail to mention these numbers. The Australian Koala Foundation estimates that less than 100,000 Koalas are left in the wild, possibly as few as 43,000, the lowball number I used earlier. However, the link to a page on how this estimate came about was broken. Let's go to the science literature to get a more robust estimate.

Phillips (2000) reports koala population trends in the reputable science journal *Conservation Biology* as declining from a population size somewhere <100,000 to "*an order of magnitude larger*". That's a discrepancy estimate of around 900,000 that, even for an error estimate, is quite a few more than the 43,000 the conservationists would have us believe are currently extant.

> Phillips, S. S. (2000). Population trends and the koala conservation debate. *Conservation Biology*, *14*(3), 650-659.

More recently, McAlpine et al. (2015) looked at '*Conserving koalas: A review of the contrasting regional trends, outlooks and policy challenges*' in the equally reputable science journal *Biological Conservation* and presented some density data but no actual numbers for the total population.

McAlpine, C., Lunney, D., Melzer, A., Menkhorst, P., Phillips, S., Phalen, D., ... & Close, R. (2015). Conserving koalas: A review of the contrasting regional trends, outlooks and policy challenges. *Biological Conservation*, *192*, 226-236.

I could go on, but I suspect you are getting the message. The suitable, irrefutable and relevant facts are difficult to find or may not exist. In the case of koala numbers, nobody knows how many there are in Australia right now, let alone historically. The key facts are a known unknown (although nobody close to the problem would admit to this).

In the absence of accessible facts, we might choose to rely on the Commonwealth Scientific Committee, the independent body of scientists tasked with deciding if a species or habitat type is at risk of extinction across Australia. These scientists gather information to match the status of species and habitats believed to be at risk of extinction and evaluate this information against agreed IUCN criteria, the global standard for risk of extinction.

What the Scientific Committee said about the koala in the *Koala Population Factsheet* was very little about the koala populations, that is, how many there are now or were around in the past. There's a nice map of where they are known to occur, and if you are a species with a distribution that covers a third of a continent, you are probably not too worried about immediate extinction—compare that to the ghost orchid known from a few tiny sites. But even in the absence of evidence and truth that the koala is unlikely to go extinct soon, if at all, the federal environment minister listed Australia's most 'at risk' koala populations in April 2012 as vulnerable under national environment law.

This means that the system set up to evaluate the evidence for decision-making on whether a species is at risk of extinction in the wild did not find enough evidence, but it made a decision anyway. Perhaps the scientific committee invoked the precautionary principle or temporarily put objectivity aside in favour of political expediency or public opinion. They wouldn't, would they?

What to do?

The koala extinction example is typical. It shows that obtaining and evaluating enough relevant facts can take a lot of work to be truly objective. There will usually be holes in your own or the collective knowledge, known unknowns and unknown unknowns. It is rare that all the relevant facts exist or are accessible. Yet we will be told that the koala is in danger of extinction; therefore, we must do all we can to protect it. Most of us don't have the time, inclination or access to the facts to evaluate this statement objectively. You can find scientific references on Google Scholar, but you also need to know how to look, for not all sources are credible. It takes too much time to hunt down and assemble the facts before you put in the intellectual effort to evaluate them and find the truth.

We would also have to work hard on ourselves to suspend our emotional response. And we already know why koalas, pandas, polar bears and other creatures with fur are used in these extinction statements—to humans; they look cute. Given that all this searching, accessing and evaluating is hard work, we resort to our default source of objectivity: what we feel. Our instinct, the gut response somewhere deep down, is part guide and part conscience. Should we be brave enough to discuss our objectivity with others, this gut feeling will be exposed to severe peer pressure and must pass some collective logic from group gut feel, known around these parts as the pub test. No, all this effort is too much. It is easier to give in to the cuteness and drop $10 into the 'Save the Koala' collection tin.

Koalas are going extinct, just one of the many opinionated statements we must filter daily. And the filter is hard to apply. Koalas are cute because human beings have a genetic predisposition to find expressive forward-pointing eyes on either side of a nose clustered together relatively low in a much more prominent face impossibly attractive... Think about it.

We will need many compelling facts to overturn such innate prejudice. Most people would not even see the need to try. Koalas are cute; what's your problem, dude?

Key points

- NRMA, an Australian insurance company, promised to plant trees for each new home insurance policy to ensure the future of koalas, an implausible claim due to the scale of the initiative versus the actual impact on koala survival.

- Determining the actual number of koalas in Australia is challenging due to their elusive nature and the variability in research estimates, making it difficult to ascertain the extent of their decline or the level of threat they face.

- Koala conservation shows the difficulty in maintaining objectivity in assessing issues like extinction, often influenced by emotional responses and societal perceptions about the cuteness of certain species. Inherent human bias towards cute or appealing animals, like koalas, can influence perceptions and hinder objective evaluation regarding their conservation status.

- Objectivity requires rigorous evaluation of facts and evidence, challenging ingrained perceptions, and delving beyond emotional responses to understand the reality of conservation efforts and species preservation.

In the next chapter

Now we have confirmed what we knew instinctively, that koalas are cute. What does it mean for how we think about broader environmental issues? Let's explore the bigger picture.

Chapter Three

How valuable is the planet to you?

People care about cute animals, the Pope and the Denver Broncos. And the earth, too, someway down the list of priorities.

Care for the earth

We are all immediate creatures living in and of the moment as we try our best to provide for ourselves and those we care about. Our ability to perceive the planet is limited even in this age of digital technology, Google Earth and Sir David Attenborough. We know our backyard, at least the parts of it that we frequent. But if we live in the city and get food from the supermarket and water from the tap, we have little idea where those fundamentals for life come from or how they are produced. Extending our perception out to the planet has us all struggling.

It is surprising how few modern, supposedly well-educated youngsters know that Africa is a continent and not a country or that the population of Nigeria is 215 million, meaning that for every three Americans, there are two Nigerians. It's a stretch to know that the five most prominent cities in China are home to 80 million people. This is twice the number of people living in California, but you get the point. Generally, we are ignorant of the planet and only know little about where we live. This might explain some recent survey results.

Actions to preserve the planet

When surveys ask people the priority actions to save the planet, their typical answer is... **what I am doing already.** Respondents were proud of what they were already doing (74%), said that the experts couldn't agree on the best solutions (72%), and needed more support from the government before they could act (69%). They say that the important things are those established habits that take little effort or come with no responsibility. People also say that they can't afford to do more (60%), they lack information (55%), and they say that individual efforts don't have an impact (39%).

There is less support when it comes to taking action to help the planet, which affects lifestyle choices. One in three becomes one in four respondents who would reduce energy consumption (32%), use public transport over cars (25%), radically change agriculture (24%), ban fossil fuel vehicles (22%), reduce meat consumption (18%) or reduce international trade (17%). Along with this aversion to taking action, a third of respondents say environmental threats are overestimated (35%), and a third don't have the bandwidth to think about it (33%). Here is a summary quote.

> Asked which actions to preserve the planet should be prioritised, moreover, people attributed more importance to measures that were already established habits, required

> less individual effort, or for which they bore little direct responsibility.
>
> Jon Henley, Guardian journalist.

At the individual level, most people tell themselves they are on it and do not need to do anything different. So, the following headline stat is surprising—71% of respondents would prioritise environmental protection.

Observe politics in any capitalist economy, and it will not be long before some politician or commentator mentions jobs, their most frequently uttered word after growth. The Canadians even had a *Jobs and Growth Act* passed in December 2012 by the Conservative government to implement its budget. The way jobs are discussed, you would think it's everyone's preoccupation. And why not? Everyone needs employment to progress, and life is challenging at best without a job. So the evidence of public opinion, summarised in a recent report by the Pew Research Center in the US, should shock our parliamentary friends. Surveys of around 1,500 people representative of the general public in each of 20 countries with mature or emerging economies across Europe, the Americas, and the Asia-Pacific suggest that **71% of respondents would prioritise environmental protection. In contrast, a quarter would prioritise job creation**.

Yep, hard to believe. A clear majority of the people in mature economies are concerned about the environment. Their intuition is telling them something that politicians don't seem to get. Not only have public priorities on environmental protection risen over time, but most people across all countries say they would prioritise protecting the environment even if it causes slower economic growth. Not a 50:50 split, the kind that pollsters say is too close to call, but three out of four people have got the message that there is an environmental challenge ahead of us and that something needs to be done about it, even at the expense of job creation. Here is the country breakdown from the Pew Research Centre.

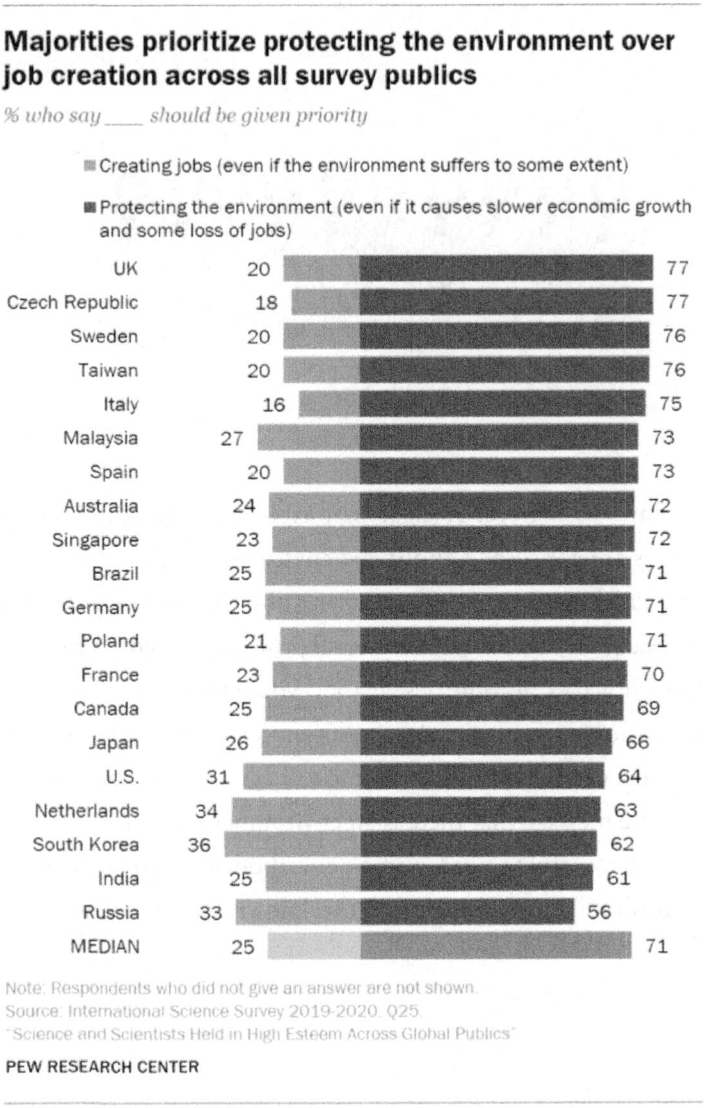

Now, perhaps the 70% plus in favour are already in jobs. The average unemployment rate in mature economies is usually less than 10%, meaning that only 10 in 100 people are out of work. It is easy to say look after the environment when you are in a stable job. The results might have differed if the survey had been completed immediately after the COVID crisis shocked employment. Nevertheless, the rhetoric from politicians always talks about jobs, jobs, and more jobs through economic growth. They suppose this taps feelings of insecurity. Only it seems people are also worried about a rupture of the environment. So, in the spirit of this zigzag contradiction, how come this happens? A topic from left field might help with an answer.

Why not ban an insecticide that kills bees?

Thiamethoxam is an insecticide, a neonicotinoid chemical with a broad spectrum of activity against insects. It was approved in the US in 1999 as an antimicrobial pesticide, wood preservative and insecticide and is still approved for use in various crops. The Food and Agriculture Organization (FAO) of the United Nations rates thiamethoxam as "*moderately hazardous to humans (WHO class III)*", non-toxic to fish, daphnia and algae, mildly toxic for birds, highly toxic to midges and acutely toxic for bees. Wait a minute, **acutely toxic to bees**. Yes, bees. You know, the insects that

- are essential for pollination of kiwifruit, Brazil nut, watermelon, melon, squash, pumpkin, gourd, marrow, zucchini, macadamia, passion fruit and rowanberry.
- are the dominant pollinators for 26 other crops, including apples, avocados, pears, mangos, and numerous berries.
- as Forbes reported in 2019, contributed between $235 billion and $577 billion worth of annual global food production. The

global recorded music industry revenue was $26 billion in 2021.

Why would we use an insecticide that kills animals essential to our food supply? Well, now we don't. In 2018, the European Commission banned the outdoor use of thiamethoxam because of the potential to harm bees. Growers can only use the chemicals in permanent greenhouses, except where we do. The US Environmental Protection Agency proposed an interim decision to allow thiamethoxam use, but with new measures to reduce risks to pollinators and protect public health. Measures included additional PPE for farm workers who handle the pesticide and restrictions on applying the pesticides to blooming crops to protect bees and other pollinators. The UK government authorised thiamethoxam to be used on sugar beets in England in 2022 because of the potential risk of yellow viruses spread by aphids, which could severely damage 70% of the national sugar beet crop. Note the word 'could' and that Brexit has benefits after all.

As Craig Bennett, the chief executive of The Wildlife Trusts in the UK, pointed out,

> *Government has outlined ambitions to restore nature, promising to protect 30% of land by 2030 and reverse declines of precious wildlife—but at the same time, it is giving a green light to use a highly toxic chemical that could harm pollinating insects and pollute soils and rivers.*

Yes, it all seems a tad contradictory. But wait.

The UK National Farmers' Union (NFU) says 3,000 farmers grow sugar beet, and the wider industry supports around 9,500 jobs in England, mainly in the east of the country. What if the farmers and related industry workers were compensated for any crop losses instead of allowing a damaging pesticide? Suppose that means 10,000 jobs at the average wage, roughly $42,000, are impacted for a year—cost: $420 million. The economic value of insect pollination in the UK has been estimated at between $485 million and $813 million. So there you have it. The future loss of production is less significant than the costs of saving jobs today. Governments know the importance of keeping people gainfully employed.

So why not ban an insecticide that kills bees? It's politically expedient not to. Politics gets in the way, even though bees are critically important to agriculture and should be protected from harm and their habitat enhanced. And no, micro-drones are not the answer. Bees are also a metaphor for how little we appreciate nature's interconnectedness and dependence on it. What happens too often is that decisions are emotional and fueled by a desire rather than objectivity. It is almost trite to say so, but the following example shows how far the media is prepared to go for the eyeballs and the clicks.

Destruction, destroyed and other emotive words

May 2019 was election time in NSW. In one of the regional newspapers, a short article appeared during the campaign period that used the word 'bulldozed' twice and the terms 'clearing', 'destruction', and 'destroyed' three times each. The 338-word article also contained emotive words like axed, loss, ban, and excessive. And there were some choice phrases peppered throughout, including

- bulldozing of native bushland

- habitat was destroyed
- in the past year's destruction
- wiped off the map
- opening the floodgates to the destruction of forests
- devastating spike in deforestation
- just the tip of the iceberg
- matter of urgency
- destruction of habitat is accelerating
- on track to be extinct
- stop this excessive tree-clearing clearing
- destroyed habitat
- forests that were bulldozed

Any idea what the article was about? Remember, it was written, or perhaps created, just before an election, and at these times, people or generative AI engines can be prone to a bit of hyperbole. Well, here is the article headline.

> *Koala habitat destruction nearly doubles in North West since repeal of native vegetation laws*
>
> Moree Champion

I know, you guessed it, the koala is back. The devastating floodgates of excess have destroyed this cute and cuddly animal with bulldozers. I added the cute bit. Except this is a lie. Of course, it is. The article was most likely generated by a bot from a few seed words, even though it claims to be quotes from real people. As we have established, the truth is people are still determining how many koalas there are in NSW; not even the science specialists have any idea, and there is no obvious way of finding out anytime soon. Carefully avoiding the evidence, the appeal is a blatant play on emotions.

Why not run with evidence that the koala is an adaptable and persistent species that benefits from human intervention in the landscape? And because koalas gain nutrition from young eucalyptus leaves, vegetation regrowth provides higher quality food sources than mature forests. Or mention that the species has a wide distribution and is persistent in some remote or inaccessible habitats. Or that it is cryptic, meaning it is hard to locate and is generally underreported even from high-quality habitats. Bizarrely, the best way to find them is with dogs.

Then, suppose all the emotional hyperbole wasn't enough. In that case, the issue of habitat loss that the article purports to expose, which for koalas is the removal of trees, is a genuine threat to local koala populations and not the species as a whole. We call it land clearing and have been doing it for generations. It is real and has nothing to do with koalas. It is all about land value. Land with trees removed in the region the rural newspaper reports is in the northwest of NSW goes for twice the value per hectare than land still covered in native vegetation. Should a farmer apply for and receive permission to clear the vegetation from 100 hectares of his land, his property value doubles and he can take that extra equity to the bank for a loan or sell up and retire to a coastal villa.

What the mindful sceptic sniffs is a rodent. All the destruction, devastation and tracks to extinction are designed to get us fired up about what happened here in this tiny patch where a koala lived so that we don't ask about what happened over there, there, and there. Follow the money is a handy adage for crime shows and investigative journalism that applies to the stories about saving the koala. If obfuscation is the best you can do, then God must help the koala because this nonsense will not.

Here are some more facts relevant to the koala case.

- Species go extinct. That is how evolution works. As species go extinct so, new species evolve for a net gain in diversity over evolutionary time.
- The koala is cute.
- We do not know how soon the koala will go extinct. We do know that it will, eventually.
- The koala is an iconic species feted by overseas tourists who value seeing this species as a big part of their coming to Australia. This tourist revenue generates significant employment.
- Humans have cleared land for agriculture for 12,000 years.

I could go on. The point is that the evidence is very different from the rhetoric. That the normative words are believable has more to do with the emotion they invoke than any of the facts. Knowing this is what gives the mindful sceptic the advantage. She knows the ropes, knows the emotional snares and seeks the truth. She becomes a pragmatist by applying critical thinking to become objective. For the mindful sceptic, the facts must be the facts, not opinions, emotional triggers, or slights of hand.

Key points

- People prioritise personal well-being, family, and local communities over broader environmental issues, partly because of human nature and limited bandwidth for caring.
- Emotive Language and rhetoric in media and political discourse can manipulate public perception of environmental issues through fear and emotional responses, often clouding judgments with emotions.
- Shifting priorities demands significant effort and sacrifice, as individuals are naturally inclined to prioritise self-preservation and immediate concerns even as they claim to support environmental protection over job creation, contrary to political rhetoric.

In the next chapter

An assumption has gone unnoticed. I didn't mention it, but you made it—extinction is bad. Before we go any further, we have to look more closely at extinction.

Chapter Four

Emotions from actions

Grazing animals find it hard to stop reproducing. There are at least ten tree species for every mammal species on the planet. Politicians would like a zero extinction target. And extinction is forever.

Black rhinos

I took this picture in 1988. The building is part of the Mana Pools National Park warden's complex in the Zambezi Valley, Zimbabwe. The skulls of black rhinos were collected by Wildlife staff as they waged an unwinnable war against poachers. At the time, it didn't seem that anything could be done to save rhinos from exploitation. Rhino horn from one animal provided enough income to feed a poacher and his family for a year, which meant a lot. Plus, wherever there is a resource, humans seem to find a way of exploiting it for commercial gain. Ultimately, the Zimbabwe government found it too costly to police the remote riverfront environment, and the rhino population continued to decline. By 1993, the total number of black rhinos in Africa was 2,400, and the Zambezi Valley, formerly the largest stronghold of the species, had just a handful of individuals.

A translocation program was initiated when it became clear that prevention wouldn't work. Rhinos were darted and, much to their annoyance, transported south to ranches and reserves further south and across the border into South Africa. On one visit to Mana Pools, we met with some of the staff on the translocation project, and I stood next to a boma made of sturdy mopane poles just big enough for a rhinoceros. Inside, an angry young bull had given himself a bloody nose, testing the strength of the construction. I climbed to the upper rungs for a better view, and there I was in another surreal Africa moment, having dominion over a majestic beast scared out of his wits.

At the time, translocation seemed a desperate, costly, and probably impractical approach. Yet, despite the ongoing poaching that even today sees over 500 individuals lost per year, the overall population numbers began to increase. By 2008, the total population of Black Rhino in Africa was 4,240, and in 2021, there were over 6,000 individuals, most of them in South Africa. The IUCN still classifies the Black Rhino as Critically Endangered, but it is in better shape than it was a few decades ago and certainly much better than when I climbed the boma.

Here is the thing. That population trough I witnessed in Zimbabwe in the late 1980s was down from estimated numbers of over 100,000 when I was born. Here is the full graph from Our World in Data.

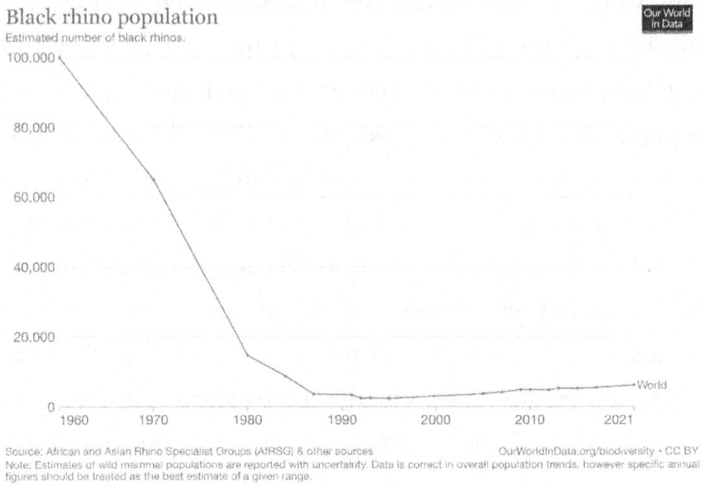

Dominion can be cruel. The Black Rhinoceros is just one of the many species with a history of decline. The numbers are accurate, along with the acute risk of extinction from pressures that are hard to stop. The only option if the objective is to avoid an extinction outcome is intervention with active management. Effort, commitment and resources are engaged and must be maintained indefinitely, and there is no guarantee of success. Translocation and breeding programs on ranches and protected reserves did work for *Diceros bicornis.* The WWF calls the recovery in numbers "*a tremendous comeback from the brink of extinction*" before recognising that "*a lot of work remains to bring their population up to even a fraction of what it once was—and to ensure that it stays there*". Once active management is invoked, it usually has to stay. Whilst few of the remaining black rhinos are genuinely wild, for now, thankfully, they still exist.

Kangaroos

Sometimes, dominion has unintended wildlife consequences in the other direction.

> *"If they're plague proportion, well, they do something about it, which is understandable, because they have to make a living, same as everyone else. But a lot of properties, in my eyes, they're not in plague proportion. And they cull kangaroos for the sake of doing it."*
>
> Catherine Cusack, Member NSW Legislative Council

Nothing out of the ordinary here. There is and always will be a visceral response by some to the removal of wildlife, be that elephants in Namibia or baby-faced seals in the Arctic—the c-word is heavy with intent. Even the formal definitions of the verb **to cull** are emotive, '*to reduce the population of (a wild animal) by selective slaughter*', '*to reduce or control the size of (something, such as a herd) by removal (as by hunting or slaughter) of especially weak or sick individuals*'

In Australia, a Liberal party member is approximately equivalent to a Conservative in the UK and a Republican in the US. Right of centre in politics is usually about resource use, liberty for individuals to choose what to do with resources, and a government that keeps its nose out of such business wherever possible. It is rare for the right to criticise primary producers on conservation grounds. They are more often championing their right to farm. It makes the quote above from the Liberal Party MP Catherine Cusack unusual.

Maybe Ms Cusack likes kangaroos, given she grew up on a farm in rural NSW. She was demoted in her political party when she crossed the floor to vote down legislation to protect koala populations in NSW because she said it had '*zero to do with protecting koalas*'. True, as it happens. These are strong green credentials for a representative of a party that has become colour-blind when it comes to nature. But as so often happens, the emotion gets in the pragmatist's way. So what's with kangaroos, the odd-looking herbivores of the Australian outback, that makes some farmers feel the need to cull their numbers?

Kangaroos have few natural predators. Dingos and Wedge-tailed Eagles will take young, as do introduced carnivores, such as wild dogs and foxes. The thylacine, considered by palaeontologists to have once been a major natural predator of the kangaroo, is now extinct, as is the marsupial lion; also gone are the *Megalania* (a giant monitor lizard) and the *Wonambi* (a sizable snake) that were likely predators. Adult kangaroos still take flight en masse at the slightest disturbance, but these days, they only have to worry about humans.

Kangaroos compete with introduced herbivores, but the critical resource is water. Access to reliable water, especially during drought, is the most likely limit on kangaroo numbers. Give them reliable water, and their population grows, often to the extreme. In reality, kangaroo abundance in many areas is high because humans have added water sources for their livestock across the arid landscape. No matter if kangaroos are soft-footed, they still eat. And this is what the farmers object to. So would you if your hard-earned went into the stomach of a herbivore that seemed to do very little other than make more herbivores. Many farmers see the need to control the number of animals they know can become a pest.

At the same time, we are asked for a donation to save the koala and must spend millions to translocate megaherbivores; there is a need to cull other native species. What is going on?

There is no such thing as passive management

Actions always have reactions, and with kangaroos, adding water to the landscape allows a water-limited animal to prosper and, along with the livestock, impacts the vegetation over and above what would have happened without water points. It doesn't take much for the grazing pressure from livestock and the added attention of prolific kangaroos to result in overgrazing that degrades the vegetation and soil. Short of changing the climate to a wetter one, the only way to recover degraded soil is to rest the vegetation from herbivores, followed by moderate, preferably pulsed, grazing. What gets called a cull is not about a plague animal but about vegetation and soil. It is called active wildlife management, but really, it is ecology. If kangaroo culling is not palatable, then something must be done with the water. That cannot happen, though, for the sheep and cattle have to drink, too. And the farmer has to make a living. Active management is the only way to accommodate multiple values.

What the mindful sceptic knows from following this tale of emotion, economics, and ecology is that the balance of nature is precarious and easily tipped over. Moreover, it is hard to prevent a fall, and constant effort and action are needed to maintain a desired state of nature once the natural balance is upset. Active management for some goal, target or desired environmental condition is complex, costly and endless.

Active wildlife management

Human interference in the natural order of things always has consequences. When humans were hunters and gatherers, they changed their habitats and landscapes soon after they arrived. They ate the easy-to-catch animals, burnt the vegetation, altered the flow of streams and moved plants around. Even the rainforests that look so natural are influenced by generations of humans planting trees and managing gaps for the regrowth of preferred plant species.

When Catherine Cusack was questioning the motives of some NSW farmers, a report from a parliamentary inquiry in neighbouring South

Australia recommended that the state's environment minister decide to declare koalas, western grey kangaroos, long-nosed fur seals and little corellas overabundant in some areas. Such a declaration would allow active management, including a culling option. However, the report acknowledged a reluctance to communicate the need for culling publicly because '*some community stakeholders find the concept of culling an abhorrent approach in managing overabundant species*'.

Although we do, there is little point in getting hung up about culling. It is just one of the active wildlife management options necessary once humans begin exploiting nature. Admittedly, it is brutal and relies on a '*long-term good*' logic that can sometimes be hard to believe when our lives are used to the immediate. But it is, at times, a necessary tool. What is more important is the misunderstanding in the MP's quote when she says, '*for the sake of doing it*'. Few farmers will go out with a shotgun and cull meaningful numbers of kangaroos or hire professional shooters to do the job for them without a reason. They believe the animals are stealing fodder from their sheep, degrading the soil, or both. They cull numbers and not just take out a few animals for lean cuisine. They do it to manage the land.

The real point should not be to belittle their actions but to be more informed about when such extreme intervention is necessary for the long-term productivity of a landscape with kangaroos. Active wildlife management is challenging in practice and intent.

How about we rest from the wickedness of mammal management and ask conservation questions about trees? Surely, they are easier to handle.

How many tree species are there on earth?

A review of the taxonomic literature where specialist scientists formally described species counted 6,495 species of currently recognised mammals (96 recently extinct, 6,399 extant). This was an increase of 1,079 species in about 13 years thanks to the efforts of taxonomists in making new descriptions at a rate of ~25 new species per year.

Hold on. An **increase** in mammal species? Everyone has told you, including in the *Mindful Sceptic Guide to Biodiversity Loss*, that a biodiversity crisis is happening, and the sixth mass extinction is in our faces and on our hands. Well, yes. Current extinction rates are way above the baseline recorded from fossil records and genetic data. Still, the taxonomists are few and have a mountain of classification work because not all species are described by science. We do not have a definitive catalogue of all life on Earth, an indictment of our care and due diligence levels. So here is a question. Do you know how many tree species there are in the world?

58,496 species, according to a recent report on *The State of the World's Trees* produced by the environmental NGO Botanic Gardens Conservation International. This number is an order of magnitude larger than the number of mammal species. We should not be surprised.

Trees have been around for nearly 400 million years, twice as long as the mammal lineage, giving them plenty of time to become more diverse. Time and disturbance are all natural selection needs to create endless possibilities. The report also says that at least 30% of the world's tree species face extinction in the wild. Well-known oaks and magnolias to tropical timber trees are among the 17,500 tree species at risk—twice the number of threatened mammals, birds, amphibians and reptiles combined. As with the koala, we must examine the evidence for these claims.

How is tree diversity and risk calculated?

Research for the report collated information from the IUCN Red List Categories and Criteria (IUCN Red List, 2021), the most widely used system to assess species' extinction probability. The IUCN Red List uses standardised assessment procedures to assign species to different extinction risk categories based on five quantitative criteria, including measures of population sizes, restricted geographic distribution and rate of decline. Assessments are also complemented with a map and additional supporting information, including specific threats, uses and ecology. The information for these assessments comes from the research literature where tree taxonomists, ecologists and foresters report on surveys of trees. Here is how tree species split among the IUCN categories.

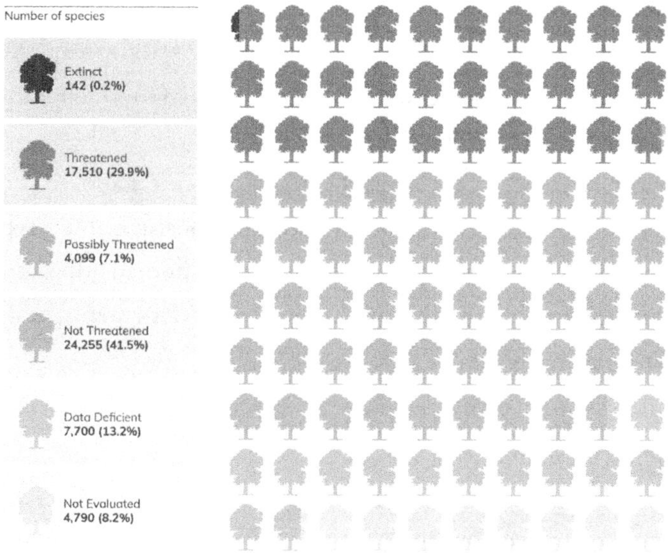

Source: BGCI (2021). State of the World's Trees. BGCI, Richmond, UK

You might notice a couple of things from this graphic beyond the high proportion of species categorised as threatened. A fifth of all tree species are data deficient or not evaluated (21%), meaning that scientists do not know enough about the trees on the planet to make a precise call on their population status or trends. We are talking about trees—significant organisms that do not move around and can only hide by looking like another tree. Humans have used trees for millennia and know a lot about how and where they grow, so not knowing the status of one in five tree species on Earth is fascinating.

The second thing is that almost half of all species either do well for numbers and distribution or are only possibly at risk (49%). It turns out that this is true for most organisms. Each group has common species that are common for a reason and likely to persist irrespective of human activity, but I digress. We are on trees. The biggest threats to tree species globally are forest clearance for crops (impacting 29% of species), logging (27%), clearance for livestock grazing or farming (14%), clearance for development (13%) and fire (13%).

In short, tree species are at risk from people—the standard problem for conservation. Climate change might be on the list, but it is typically a confounding or accelerating factor added to the human impacts. A tree cannot run and has trouble hiding in the days of GPS systems, chainsaws and transport machinery.

Reliability of evidence

Whenever a number is an answer to a question, it begs another question. How reliable is that number? 58,496 is not the number of tree species on Earth, even though it is presented as precise. It estimates the actual number because not all tree species are described or all habitats are surveyed. As species are found from descriptions or survey discovery, the number goes up and down with extinction events and sometimes with taxonomic revisions. So, for most purposes, the number itself is far less critical than how reliable that estimate is and if the number is trending over time.

A mindful sceptic is wary and follows the recommendation **never to leave a number alone.**

Is 58,496 a significant number? It could be. If 58,496 were the number of children who died from preventable diseases in the last month, it would be both big and tragic and unacceptable. If it were passengers aloft in commercial airliners at any given time, then a whole heap of airlines would go bust overnight, given there are roughly a million people in the air at any one time. If it were potholes per kilometre of road, you would be in Zambia.

The problem is that 58,496 needs more context and comparison. On its own, it isn't significant. I could tell you that a similar number, 90,000 is the area in hectares of native vegetation cleared for agriculture as a policy review trigger in NSW. Farmers in the state can apply to manage or clear native vegetation on their properties legally, and when the cumulative area of approvals reaches 90,000 ha, it is agreed in advance to take a close look at the policy that allows clearing. Noting that during the policy development, any agreement on the trigger number was hard-fought, for there are advocates for zero hectares and those who believe that no number is big enough. At least this 90,000 has context. It represents an area of land and has a purpose. Even with context, 90,000 will have advocates claiming economic progress from agricultural development and detractors who lament the loss of native plants from the landscape. On its own, 90,000 has garnered opinions, even as a trigger number. What 90,000 or 58,496 need is some company.

For the mindful sceptic to think critically, the 90,000 must have other numbers. For example, the area of native vegetation under conservation, restoration or active management (12,863,450 ha), the area of winter crop for 2018 (3,100,000 ha) or a string of numbers such as the area converted to arable agriculture each year for the last 50 years (curiously this value is difficult to pin down). Only when 90,000 ha has other numbers can it find itself, make sense and contribute to society. This is true of almost all reported numbers.

90,000 ha is a precautionary number in the context of an area treated under a native vegetation policy. As a proportion of the area of NSW (80.9 million ha), it is minuscule; even as a proportion of the area under arable production, it is small. Then, there was an unprecedented wildfire season in 2019, and over 3.6 million hectares were burnt by bushfires that raged for months, the largest covering more than 800,000 hectares of continuous forest. This time, 90,000 ha has a very different context. Fire is not the same as clearing, for most trees will recover from fire, and seedlings will be established in the ash beds when it rains, but the point is that the area needs context. Is it worth discussing if the trigger is just 10% of the area of a single fire?

Comparison is always critical when dealing with numbers. A single number often makes no sense; it's naked, self-conscious and insecure. It needs some context for clothing and some friends to compare against. Next time there is an argument over a single number—over the 3.6 million hectares of bushland burnt in 2019—remember you can't leave it alone. So, is 58,496 a significant number?

It is an order of magnitude larger than the 6,495 species of currently recognised mammals. The Smithsonian Institute reports that at least 900,000 different kinds of living insects are known. At the same time, the UK's Royal Entomological Society suggests there may be as many as 10 million insect species on Earth, with approximately 1.4 billion insects for every person on Earth, and the total weight of all the insects is about 70 times more than all the people. More significant than some, smaller than others is always the trite answer.

17,510 tree species threatened with extinction based on realistic assessment criteria does sound like a large number. Extrapolate this to insects—hard because the evidence for insect numbers is patchy at best—and there could be 269,100 known insect species at risk of extinction or nearly 3 million species if the complete inventory were available. The number of koala species that may or may not be at risk of extinction is one.

Zero Extinction Target in National Parks

One consequence of a small number focus is odd policy options like this one. The state government of NSW, Australia, responsible for the people living in a land area of over 80 million hectares, announced zero extinction in national parks as a policy target. Zero extinction is a great slogan. It sings of hope and justification and has a 'vote for me' vibe. It also feeds the unique human requirement for dominion—our job is to lord it over nature. And we are so good at this control that we even believe we can prevent extinction. If we were around when the dinosaurs copped it—we weren't—we would have saved them. Later, I'll return to the naive belief in our god-like capability, but first, let's look at the zero extinction target.

Resolutions for the future

Each new year, many of us set a resolution for the future. We put ourselves in the frame to lose weight, get fitter, read more books, save money, or achieve many life-affirming targets. It is often a long list of self-help aspirations. But the descriptors—lose, get, more, save—are all relative. They can only apply in context, as does any numerical measure of aspiration or goal. If I resolve to get to 81 kg body weight on the bathroom scales when I get up in the morning, there is no information on the likelihood of achieving such a target in the next 12 months.

If I am a 6ft 6 professional basketballer or a 5ft 10 clinically obese 55-year-old with a heart condition and a beer belly, then 81 kg is impossible or a stretch target at best. In the United States, 36% of adults are obese, and another 33% are overweight, so plenty of people are in the latter stretch category. Luckily, I am reasonably healthy for my late middle age and currently tip the scales at 85 kg, so there is a realistic chance that I could shed 4 kg in a year. A resolution to achieve a weight loss of 4 kg sometime in the future is achievable for me, even if I have admired that target for a long time, typically from a distance. It is sensible to resolve. I am motivated to diet, fast and exercise more. And because the target is achievable, I am not messing with my head. Ask a 100 kg 55-year-old to get to 81 kg, and he will likely laugh and pour another beer.

A realistic target

A target of zero extinction of native wildlife in the NSW National Parks estate is the first time an Australian jurisdiction has set such a goal.

> *"Just as we have a net-zero emissions target, we now also have a target of zero extinctions for our national parks, and are aiming to improve and stabilise the on-park trajectory of threatened species by 2030..."*
>
> Matt Kean, NSW Environment Minister, September 2021

This declaration made many conservationists happy—one of the few times a zero-target can win votes for a politician. The problem is the context. The NSW National Park estate—called a national park even though these areas are gazetted and administered by the state, not the federal government—includes 880 parks and reserves, covering over 7 million hectares of land, roughly 9% of the land area in NSW. The good news for a zero extinction target is that another 80 million ha of land across the state exists, where wildlife species can hang out and repopulate the parks.

7 million ha is a lot of land, the same area as Ireland and a little more than Sri Lanka, a land mass big enough to have 16 species of endemic mammals. This is enough land, then. Plus, the 880 parcels of land are in a matrix of primarily agricultural land, including rangeland and native forests where some species living in the parks can also persist. Roughly 44 million ha of NSW is grazing land operated by sheep or cattle farmers on altered vegetation but relatively close to native—fewer trees but plenty of native plant species among the grasses, forbs, and shrubs. If any politician wanted to set a zero extinction target, then the NSW environment minister is better placed than most.

Only zero extinction is impossible.

There has been no point in the 3.8 billion years of simple cells (prokaryotes), 3 billion years of photosynthesis, 2 billion years of complex cells (eukaryotes) or 1 billion years of multicellular life, a time with zero extinctions. It has never happened before because nature accepts losses as part of the process that allows gain—it is how evolution works. Extinction is inevitable. Isolating parcels of land selected randomly from a much larger contiguous area and deciding that no extinction will happen in these parcels is bizarre, naive, reckless, ignorant—and a host of other descriptors of stupidity. Here is the main reason.

Extinction is inevitable

Palaeontologists have a fine time of it. They fossick around in obscure parts of the world where few people are and take a hammer to rocks. They have done this for generations, found many fossils, and came up with some interesting conclusions. Here are five of them.

- 444 million years ago, 86% of species were lost.
- 375 million years ago, 75% of species were lost
- 251 million years ago, 96% of species were lost
- 200 million years ago, 80% of species were lost
- 66 million years ago, 76% of species were lost.

These are, of course, the famous five mass extinction events in the fossil record. The last one is the most familiar, when all the dinosaurs copped it after an asteroid collided with the earth. Here is another pattern the palaeontologists discovered.

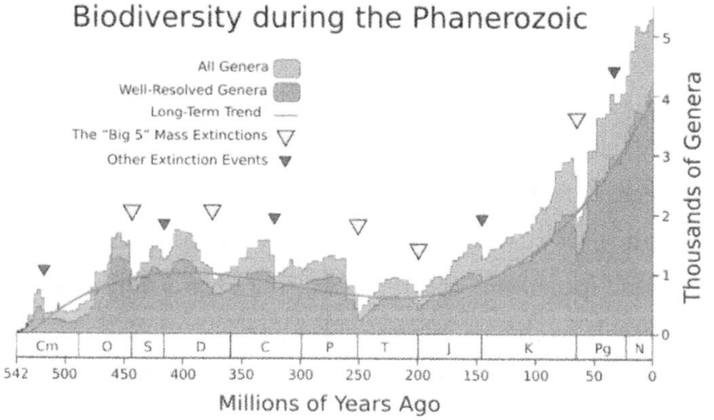

Over geological time, the trend in diversity is for it to increase—read time as right to left on the graph—despite the five big extinction events and a host of smaller ones.

Over time, diversity has **increased** despite these mass extinctions and several smaller ones in between them. More species are alive today than at any other time in evolutionary history, even though humans are kicking the sixth mass extinction into existence. This is the history of life on earth. It is dynamic, with massive shifts in diversity, but with an underlying driver powerful enough to **increase diversity over time**. If you believe this evidence, then a few core concepts become clear:

1. Extinction happens.

2. Large extinction events happen often enough to make another one inevitable.

3. After extinction events, diversity recovers.

4. For at least 150 million years, the trend has been for diversity to

increase rapidly, noting that the non-avian dinosaurs were lost during this time.

Diversity happens because there is enough time for mutation and natural selection to hone any number of specialisms. When organisms specialise, they are successful in a relatively narrow set of conditions. Some of these specialisms are inflexible, so a proportion of species become vulnerable to change in those conditions.

Enter human beings. Modestly specialised initially but with a distinct advantage. They had the brains and communication to adapt to a wide range of conditions, especially when they tamed fire and then much later discovered fossil fuels. Once in control of energy, humans have appropriated over half the global biomass production, removed vegetation, gotten rid of wildlife and altered the atmosphere's composition. There is more than enough change for a mass extinction. Humans affected the change, but a change would have happened at some point. Tectonic plates moving to cut off ocean gyres, volcanic eruptions, another meteorite strike, it would happen soon enough, maybe 50 million years hence.

Is the reality of extinction too painful?

Perhaps it is. We may be too much in the present day to accept timeframes beyond ourselves. Indeed, there will be huge consequences of extinction for our production systems, the ecosystem services we rely on, and the very liveability of the planet for our species. We'll need a lot of adaptation to survive. What we might also remember as we plan for change and seek to retain what we can of the diversity of life is that extinction is natural and inevitable. It is as much a part of the process of evolution as the creation of new types.

A zero extinction target is meaningless.

Declaring that no extinction will happen is ridiculous and an insult to the intelligence of the voters. The fossil record and modern evolutionary biology based on genetics confirm what Charles Darwin and Alfred Russel Wallace surmised: **some types of organisms fail**. They go extinct while the fitter types persist, at least for a while. Extinction is an inevitable consequence of natural selection. And because the environment is in flux with disturbances everywhere, being fitter today is no guarantee that the type you represent will be fit tomorrow or in a decade or a thousand years. A zero extinction target is pointless from the logic of evolutionary theory. Then there is the practical matter.

How, for instance, will anyone decide if one of the hundred or so vertebrate animal species we might have some data on or the thousands of invertebrate, plant and microbial species that we know very little about, have become extinct or are indeed still present in NSW parks. None of the parks and reserves in NSW have comprehensive biodiversity inventories and monitoring programs, even for the at-risk icon species. If we don't know what species are present in the parks, how can we say if any will become extinct? We cannot. Researchers and park staff cannot confirm or deny the Minister's claims.

Science cannot tell you how many species are present in this habitat because no species inventories are fully comprehensive or complete.

A zero extinction target is set anyway

As each new year ticks by, we reflect and set more resolutions. No matter whether they are relative, need a solid context or are abandoned within a few weeks. It is the process of reflection that helps. The Minister sets the zero extinction target to demonstrate that the government recognises a biodiversity loss issue and is doing something about it. At some level, there is a response to environmental degradation. This is the only saving grace of most conservation targets. They allow some reflection on the problem of how to feed and house everyone without wrecking nature.

I cannot stress how crucial such reflection can be. Nature is the source of human success. The ecosystem services nature provides are the foundation of all food production systems, and keeping them is the only chance to feed everyone well. It is a shame that this essential requirement to maintain nature is twisted to be the preservation of endangered species in national parks, but that is better than unconstrained exploitation.

As I can be cheeky at times, I am tempted to ask the minister what he might suggest be done about the water points in and around the national parks and if that is important to help achieve the target.

There is no need to take the Minister or the zero extinction target too seriously. Like most resolutions, it is the thought that counts. The real work is in figuring out how to proceed towards the target. It is easy to say I will lose weight and get to 81 kg; it is something else to work out how I will get there when my body sits stubbornly at 85 kg, no matter what I eat or don't. However, there is a more sinister undertone to zero extinction. It reeks of command and control over nature. Don't worry, the politicians say; our policy can beat back a billion years of evolutionary history because we are strong and all-powerful. This delusion of dominion is the worst bullshit. But presumably, it works for our leaders; otherwise, why do it?

Here is what I suggest as a general rule for all mindful sceptics. Next time you hear politicians come up with nonsense targets, ask them how the species that they will single-handedly save from extinction got into the national park in the first place. You can let them know the answer after they look blankly at you for a minute or two. It replaced another one.

Key Points

- Extinction is a natural and inevitable consequence of evolution.

- Diversity has increased over time, even after mass extinction events, indicating that life has a remarkable capacity for recovery and adaptation.

- When applied to national parks, a zero extinction target is unrealistic and scientifically unfounded.

- Context and comparison are essential when dealing with numerical data, as numbers alone can be misleading without proper frames of reference.

- Even if they may seem unattainable, conservation targets serve as a means of reflecting on the importance of biodiversity and the need for environmental protection.

- The notion of human dominion over nature fails to respect the processes of evolution and the interconnectedness of life on Earth. Given nature's critical role in human well-being, including ecosystem services vital for food production and maintaining a habitable planet, it is a mistake.

In the next chapter

So, the koala is not going extinct anytime soon. Despite the minister's ignorance of basic ecology and evolutionary biology, we do have to take action to avert the worst consequences of human disturbance for our survival. So what about rewilding as an idea?

Chapter Five

Is rewilding the solution to save the planet?

Rewilding a third of the planet is bold and brave, maybe even essential if humanity is to survive. But is rewilding the solution to save the environment from humanity, avoid a dystopian future and keep us within planetary limits?

Is rewilding the solution

It's not an everyday question, rewilding. Only it was asked recently on prime-time Australian television. You know I am a fan of Australian Gogglebox, where people watch TV watching people watching TV—bizarre but hilarious. Select members of the Australian viewing public at home on their couches are so funny; their wisecracks and personalities light up the screen far more than the shows they are watching. One of these programs was David Attenborough's A Life on Our Planet. All the Goggleboxers, despite different backgrounds and cultures, had fear in their eyes. Their concern over the messages was palpable. Perhaps because they believed what Sir David says, they all recognised what ecologists have known for a long time: that agriculture, fisheries and human land use have decimated wildlife populations across the planet.

Biodiversity loss should not be a surprise when just 3% of the total mammal biomass comes from wild species, over 50% of the habitable land area is used for food production, and the only true wilderness is covered with ice, tundra or sand.

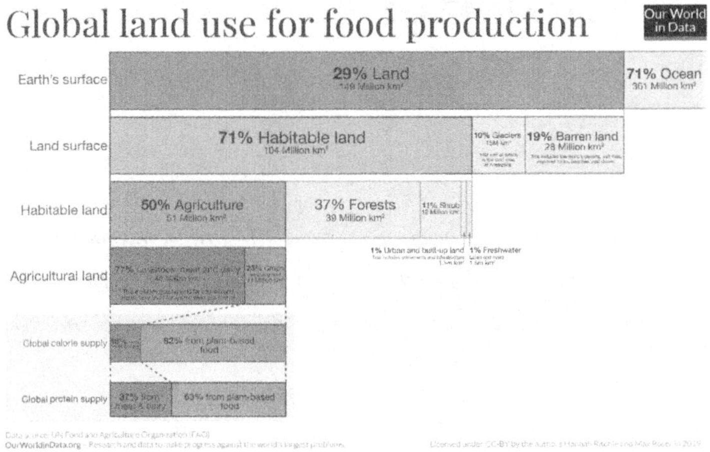

Adding climate change to land-use change means that even if we haven't physically altered habitats, the climate will do that for us. Recall that the geological record suggests that four of the five previous mass extinctions of biodiversity on the planet resulted from climate change, over millennia in those instances, but climate nonetheless. The message that Sir David delivers, and he must be commended for finally coming off the fence and telling his audience the truth after so many years of wowing them with the glory of nature as though all is well, is that **life on earth is in big trouble**.

Biodiversity, that is, all the variety of life and not just the attractive animals, is declining at rates equivalent to a mass extinction. This loss is highly significant in soils where biology drives the transfer of nutrients to plants. Sir David's proposed solution, borrowed from a growing body of scientific research, is **to rewild a third of the planet**. It is an excellent catchphrase that is attractive and accessible to many people. Return some of that wilderness that humanity has flattened to something close to its original state by finding some space for wild organisms on the planet even as humanity continues to hog most of it.

Is rewilding the solution to our mess? First, we need to know what Sir David meant by rewilding.

What is rewilding?

Rewilding is what it says on the tin. That is a return to the wild by removing human influences and letting nature take care of itself. It is nudged by the reintroduction of megafauna, such as large grazers or carnivores, to restore ecosystem processes.

More formally, researchers define rewilding as

> *"the reorganisation of biota and ecosystem processes to set an identified social–ecological system on a preferred trajectory, leading to the self- sustaining provision of ecosystem services with minimal ongoing management."*
>
> Pettorelli, N., Barlow, J., Stephens, P. A., Durant, S. M., Connor, B., Schulte to Bühne, H., ... & du Toit, J. T. (2018). Making rewilding fit for policy. Journal of Applied Ecology, 55(3), 1114-1125.

Rewilding differs from the related idea of ecological restoration, which requires a prescribed endpoint and often, as we have seen, a lot of effort in active management by being open-ended. Rewilding intervention might mirror the ecology and species compliments that were present in the past, but reorganisation proceeds according to natural processes to an uncertain endpoint or even an endpoint at all. So long as the ecosystem is resiliently providing ecosystem services and this is ongoing, rewilding has succeeded. Inactive management if you like.

Eucalyptus woodland in eastern Australia is already wild and shaped by fire events but lacks the megaherbivores that became extinct after the first humans arrived.

Rewilding is a courageous idea. It is big and bold. *The practice of returning areas of land to a wild state, including the reintroduction of animal species that are no longer naturally found there* when there are 8 billion humans also needing a home and food, is a huge call. And to suggest this happens over a third of the global land area is a dramatic reversal of what humans have done for thousands of years. As Sir David says, rewilding begins by halting the processes that are degrading the system. Turn around the activities that degrade natural capital and then put extraordinary effort into rebuilding natural habitats.

A third of the globe returned to the wild should slow rates of biodiversity loss. It will not guarantee the survival of any given species; the iconic rare and endangered will still need individual attention. However, enhanced preservation of the many will see ecosystem services maintained with follow-on effects for climate mitigation and water balance.

Feeding ourselves on a rewilded planet

There's an old saying among farmers around these parts: '*You can't be green if you're in the red*'. Suppose the human population fails to feed itself or fails to allocate enough resources to maintain the well-being of the majority of the people. In that case, the disruption and unrest that hungry people will unleash on the world will undo any effort put into conservation action.

Recall the potent matter of 22 trillion kilocalories per day—the energy from food needed to keep 8 billion people from going hungry. Throwing resources into rewilding through rebuilding nature is only possible if we're also in good shape. Human instinct is to survive, and if the food begins to run out, we will do whatever it takes. This has been the case throughout human history. Most of our conflicts are about resources of one sort or another, even if they are disguised as political agendas or megalomaniac tendencies. Sir David does temper his argument for rewilding with the requirement to grow food—some limited gathering from the rewilded vegetation. However, a mindful sceptic squirms at a few drones collecting nuts from rainforest trees. It will not be enough.

There is a 30-year window to grow and then maintain an extraordinary food supply, distributed efficiently worldwide, before the demographic dividend is achieved and there are fewer people to support. Researchers differ in their estimate of the challenge, but on our current diet trajectory, global food production must grow by roughly 2% per annum for 30 years. This is a second green revolution level of intensification and production efficiency gains. Any rewilding has to go together with food supply at levels never before seen.

Fortunately, these two things are not as incompatible as they sound. Sustainable use of natural capital is more straightforward if an ecological approach is taken. That means understanding the flow of energy, nutrients and carbon into biodiversity and primary production. There are enough linkages between nature and the production of commodities that benefit both. The solution is to think of the food production systems as closed loops running at speed dictated by the innate ecological capacity of the soil and vegetation.

Where inputs are necessary, they are used with a sensitivity to natural cycling. Think of an organic system more than an intensive one based on fertilisers and tractors. Wild areas generate **provisioning services**, such as drone harvesting the nuts from the forest; **regulating services**, such as controlling climate and disease; **supporting services**, such as nutrient cycles and oxygen production; and **cultural services**, such as spiritual and recreational benefits. When all these services are integrated into the landscape, they benefit production too.

Rewilding as part of the solution

The scope of rewilding, as proposed by Sir David, is ambitious. It is a global idea—something like the 'man on the moon' audacity of the early 1960's. And we need visionary leadership right now. Is rewilding the solution? Sure, with a caveat. Rewilding has to include a rewilding of the production system, too. Not just the spare land that we don't currently use. It has to be an integrated approach to food production that maintains ecological processes across the landscape. Rewilding just the parts that are not under production will not relieve the pressure on ecosystem services. In other words, we rewild landscapes rather than create wilderness or endless nature reserves.

We already know that the reserve systems don't do enough to halt or even slow biodiversity loss, which is part of Sir David's original lament. So, the only solution is to integrate wildlife conservation and the management of nature into the food and fibre production system, making them an integrated connected unit across landscapes, ideally the entire planet. If rewilding is to work, it needs a more extensive remit than reintroducing a few iconic herbivores and their predators. It needs to become a new term for the sustainable use of natural capital—that is, maintaining cyclical nutrient and energy processes through ecology alongside human inputs. It can be done. I would argue it has to be done before humanity eats itself out of existence, but doing it will be hard.

A mindful sceptic meanders through the challenges of rewilding 30% of the planet

Nature is resilient and will recover quickly in degraded areas given a chance. Rewilding is readily achieved by natural regeneration plus an introduction or two. All that has to happen is that human impacts are removed or mitigated. At least 30% of the land and 30% of the ocean must be hands-off so that nature can recover, degradation can slow down, and the essential services that nature provides can persist.

At its crudest, the challenge is **to rewild a third of the planet and feed everyone well**. Sharing a third of the planet with nature is reasonable, given two key assumptions: 1) that living standards continue to grow for those doing it tough, and everyone maintains their well-being with an equal opportunity for a slice of the pie, and 2) rewilding will deliver the ecosystem services needed to meet the first assumption.

Large areas are already wild, especially in the drier parts of the world.

Let's assume for a moment that it is possible to convince the majority that rewilding is essential to save humanity and that everyone is on board to achieve the 30% that the scientists say is necessary. A magic wand would be nice, but the reality is that choices and decisions must be made on various issues. Here are a few of them.

What third will go wild?

The first challenge is to decide which third of the planet is selected for rewilding and who is prepared to give up their territory to achieve this outcome. The equitable way is to have every jurisdiction give up its proportional share, so whatever territory a country claims, a third of it returns to nature. That will make little sense.

Take the Netherlands, a small country with a land area of 41,543 km^2 and a population of 17 million people. Dutch farmers make the Netherlands the world's second-biggest food producer by export value. In 2019, the country sent €94.5 billion in agricultural goods, including ornamental plants and flowers (€5.8 billion), dairy products and eggs (€4.3 billion), meat (€4.0 billion) and vegetables (€3.5 billion) to external markets. The Dutch have become incredibly efficient at producing food.

It would be madness to reduce a third of their landscape back to nature because then we would have to find equally efficient production from other areas; many times more land area would be needed because not everyone grows food as efficiently.

Conversely, some areas are much more likely to rewild faster and more effectively than others. Which countries have the good stuff, and will they accept the opportunity costs? We have excellent predictive science, so it's not a technical problem. Which third to rewild is a geopolitical problem, and humans are spectacularly poor at such cooperation for anything other than keeping the peace long enough to trade.

The Netherlands produced 952,000 metric tons of cheese in 2021 with an expectation of one billion kilograms by 2025.

2% for 30 years

The current rule of thumb for net growth in food production to feed the global population is 2% for 30 years on our current consumption trajectory. This estimate means that food availability must increase by 2% annually to meet the demands of 8 billion people, increasing in numbers at 8,000 per hour over the next 30 years.

Not all of this has to come from growing more food. There are savings in diet, in the supply chain, and especially in reducing food waste, but the reality is that food production will increase. Simplistically, reducing the agricultural area by a third cannot mean that the net amount of food production will change.

Food will come from the remaining 66% of the production land. Instead of 2% growth, that land will need to produce 4.7% more food per annum every year for 30 years. Overall, this is an increase of 140% over the current rates of production. This is unrealistic.

Already wildland

We will come to the option of changing our diet to reduce demand in a moment, but first, we need to consider what is already in use or put aside for nature. Land humans find helpful for food or fibre production is generally used for that purpose. The fertile land grows crops, the less fertile land grows livestock, and the forests are there for the timber, often on soils unsuitable for anything else. Equally, we fish in the best and the most accessible waters. The national parks, reserves, and wilderness areas are typically on land where it is hard to profit from any productive uses. On land use criteria, 29% of the land is barren or under glaciers and ice sheets. Does this count towards the rewilding, given it is already wild?

If rewilding a third of the landscape applies to the habitable land, then forests and shrubland already account for 48% of this area (34% of the total land area). If rewilding applies to 50% of the habitable land used for agriculture, then restoration of 15 million km^2 of agricultural land is required. The problem is that profit and subsistence are potent forces. When it comes to the individuals currently owning or managing the land for profit or to feed their families, there will be strong resistance unless it is evident that profit-making can continue. Where rewilding happens matters to many people, and they each hold strong values and opinions about what should be done with the land.

This landscape looks wild already but is actually managed for grazing and timber production.

Rewilding is a bold concept

Rewilding is logical and achievable in theory, but... There tends to be a problem when humans are involved because human nature gets in the way. One of the reasons humans are so successful is that we are incredibly ruthless. We will do whatever it takes to get what we want. Think of the former President of the United States chirping and weaselling his way like you wouldn't believe, to use any means possible to give himself half a chance of re-election even though he lost. And a lot of people are like that. The desire to win and not to lose is a compelling emotion.

The middle ground of win-win is the logical solution, but people prefer to be winners outright. Think of this from the point of view of a subsistence farmer in Africa or South America, where winning means food for themselves and their family. Losing their land to rewilding cannot mean the loss of livelihood. Still, it must also account for the opportunity cost of losing land, including the security that ownership gives the farmer and his descendants. The boldness of rewilding is that it must address the social challenge of land as an asset, security, and home.

It must confront and solve the compensation demanded from land given back to nature (opportunity costs) and still give a purpose to those who live on the land.

Jurisdiction-level choices for rewilding

The choice of whose third it is to rewild could be made across jurisdictions through an international agreement. For example, countries could be given an allocation target and meet it by rewilding in-country or paying other countries to put land aside. Something similar to the market mechanisms for greenhouse gas emission reduction could be developed that learned from those mistakes. Countries offering land credits would benefit from the payments and having nature close to their productive lands. Mechanisms to meet targets by trading services would be attractive in the same way that carbon trading was attractive to promote an energy transition away from fossil fuels. The rules need to be transparent. One of the problems with the carbon trading system was the rules only restricted emissions and sequestration within the jurisdiction. Emissions from economic activity that happened elsewhere did not count. For example, coal-producing nations did not have to count emissions from coal they mined and exported; the countries that burned the coal added those emissions to national accounts. Australia, an already high emitter per capita, is not required to include the 300+ million tonnes of coal it exports yearly in its national carbon accounts. The transaction rules for rewilding benefits must be carefully thought through to avoid similar perverse outcomes.

History tells us that that's hard for international agreements to achieve. They are OK with broad statements and audacious goals and targets rarely met. But it's another thing to figure out the mechanisms by which those targets could be achieved. It is also essential to ensure that the wealthy countries refrain from using the mechanism to keep the poorer countries poor. Rewilding in the poorer countries would need to be adequately compensated with investments in agricultural intensification or the provision of reliable food equity. Even supply chain access is sufficient to achieve food security and relieve the pressure on subsistence agriculture.

Remember that the wildest places on earth are either unpopulated, like Siberia, Greenland and the Antarctic, or underdeveloped, such as parts of Africa and Australia. However, the underdeveloped areas in Africa and parts of Asia are where population growth is fastest, and there is the greatest pressure on land to feed people.

> *Around 60% of the area expected to be urban by 2030 has yet to be built so the technologies used to create that infrastructure will have far-reaching social and ecological implications*
>
> Kate Raworth, Doughnut Economics

It is worth remembering that the story of the degradation of nature has chapters where international and national targets were set and never met. Most failed, and the few success stories are localised. Rewilding through jurisdictional-level mechanisms will be very challenging because it has to work in the aggregate or is not worth trying.

Nature on the doorstep

One premise that Sir David was trying to convey in *A Life on Our Planet* was that nature on your doorstep is the answer to most environmental problems. Untroubled nature close by integrates the natural habitat into human landscapes in a way that informs and improves both. The sell is that the human landscape becomes more productive and more efficient in the presence of nature. Broadly, this is the concept of nature delivering **ecosystem services**.

Ecosystem services are attributes of natural and managed systems that deliver support, provisions, regulation and cultural services to human society through a combination of environment and biology. Rewilding will help to maintain these vital services in landscapes, make them more resilient to disturbance, especially from climate change, and make them more helpful in feeding everyone well. Rewilding will bring these services closer. It's all good in theory, but It begs a question of design. Should rewilding be patchy, interspersed with agriculture and the built environment or reestablished in large swathes like a blanket over all the other land use? There are benefits to both that depend on the objective. If rewilding is about services, then an integrated landscape approach would spread benefits where needed. If it is about saving iconic species that typically need large areas of habitat, then blocks of rewilding would be more effective. And we are back to the value proposition again. What we want to achieve will determine the optimal design.

Back to nature is constrained

Once nature has been altered by human disturbance, it changes ecosystem structure, composition and function. It is also pushed off any trajectory of change it was already on. Rewilding is more complex than turning the car around and driving back to the garage, which is different from how nature works. What nature does is respond to the disturbance and react accordingly. Habitats change after a disturbance and develop as a consequence. If rewilding involves re-vegetation, we get what we call regeneration through natural succession or rehabilitation if active management is involved, not restoration to the original. Rewilding cannot return nature to its pristine past. Let's repeat that. Rewilding cannot return nature to its pristine past.

Humans are bright but still not sophisticated enough to manipulate ecological succession to achieve the same complex systems that nature developed before humanity. We need help getting rainforests to look and be as diverse as they were before human impact. That's not to say we shouldn't rewild; it is just that we need to have rewilding in our minds, which is a return to nature, and **nature will do what it wants with that process**.

Nature will regrow vegetation, return to ecosystem services, and rebalance the movement of energy and nutrients through the system. But it has to do it in its own way with organisms and species it can draw on. It can't and won't return to what it was before with mastodons, sabre-tooth cats, wolves, bears, and barbary sheep. Rewilding also implies multiple uses and a range of values, and this is not wilderness. Rewilded areas will require some human use of resources.

> *Environmental legislation has a traditional focus on in situ conservation and the preservation of historical conditions, which have favoured the implementation of conservation projects aiming to restore previously observed benchmarks*
>
> Pettorelli, N., Barlow, J., Stephens, P. A., Durant, S. M., Connor, B., Schulte to Bühne, H., ... & du Toit, J. T. (2018). Making rewilding fit for policy. Journal of Applied Ecology, 55(3), 1114-1125.

The pre-1788 problem

Unfortunately, it is a common mistake to assume that humans can restore nature to the original pre-human condition, a misnomer especially common in the conservation movement. For example, in Australia, the pre-1788 vegetation condition is used as a benchmark, an assumption of what the vegetation was like before Europeans arrived and began to alter the landscape. That benchmark is used as a rule and a measure against whether or not a habitat is common or rare or should be protected. As attractive as it sounds, there are several problems with this before-and-after approach.

First, we only have guestimates of what the landscape was like before 1788, the arrival of Europeans. Note this is not the arrival of humans because that happened at least 70,000 before this date. Vegetation cover, type and composition are only estimated. Biological inventories of today's habitats are incomplete, so we do not know what was present before modern records. It is guesswork.

Then, there is the landscape change that has happened in the last 230 years due to agricultural land use. Vegetation patterns have changed, soil nutrient levels are different, soil carbon has been lost, the climate is different, and the available plants and animals are not the same due to losses and introductions. There is no way that nature could return to what it was.

Consider the nutrient consequences of billions of sheep reared on native vegetation across outback Australia. The fleeces and meat from those animals were extracted from the habitat and sent to Europe, a net export of nutrients, especially nitrogen. Nature in the wild will be different from what it was pre-human impact. Rewilding can only be returning to the wild. It doesn't mean returning to a previously prescribed condition.

There were 66 million sheep in Australia in 2019, that's 2.6 sheep per person.

More policy problems

> *Rewilding (has) framed itself as fundamentally contrary to the assumptions about equilibrium ecosystem dynamics and cultural landscapes at the core of the EU's policy approach*
>
> Root-Bernstein, M., Gooden, J., & Boyes, A. (2018). Rewilding in practice: projects and policy. Geoforum, 97, 292-304.

Rewilding is emotional and political. Public policy is pushed towards preservation. It comes from the human condition that is attracted to stability and certainty and avoids change. Politicians leverage this by telling people that historical situations can be preserved, be that a species, a habitat or a historical building. However, changes resulting from human use of nature push ecosystems to their limits and species out of their tolerance, replacing them with novel ecosystems.

Meanwhile, agriculture is the primary human use of nature. It uses the most land and water, changes vegetation and depletes soils. It is also essential to feed everyone well. Politicians are aware of this, too.

These two policy areas, preservation wrapped up these days into **biodiversity policy** and food production or **land use and agricultural policy,** are often at odds. For example, in the European Union, the current biodiversity policy is based on a 'compositional' paradigm, predicated on preserving particular species assemblages and habitat types, often with solid legislation in member states that defines targets for species and habitat protection. Australia has a similar approach. Meanwhile, the Common Agricultural Policy (CAP) keeps marginal lands in agricultural production through agricultural support payments. Plus, 70% of payments under the CAP require that land is in '*good agricultural condition*' and free of '*ineligible features*' such as naturally regenerating scrub.

Back in NSW, the *Crown Land Management Act 2016* permits a leaseholder to occupy and use Crown land for a specified purpose, such as grazing. A Land Management Strategy submitted as part of the licence application creates the obligation that the licensed area is grazed by this plan. In other words, keep grazing. If the geopolitics wasn't challenging enough, rewilding could be at odds with land use and biodiversity policies within jurisdictions.

Rewilding also requires a massive clean up job.

Rewilding in practice

We may be able to restore certain iconic species, as has been happening in northern Europe and the US around wolves and bears, to bring those iconic species back into ecosystems. Still, those returning species live in habitats very different from what their ancestors lived in. There's nothing wrong with this at all. It just needs to be accepted as the way that rewilding will work.

> *In practice, rewilding projects are a multiplicity of things, responding to site history, social context, geography, ecological condition, and feasibility in both physical (how do we move that animal?) and legal senses (are we permitted to move that animal?)*
>
> Root-Bernstein, M., Gooden, J., & Boyes, A. (2018). Rewilding in practice: projects and policy. Geoforum, 97, 292-304.

Assuming we can let go of a return to the past, removing human activity from a region or an area and then letting nature recover will be the smartest and cheapest option for vegetation and soil to return to the wild. Nature is extraordinarily good at exploiting a resource, and left alone; organisms will come in and use whatever space is available. One of the challenges is that rewilding must be as fast as possible. The premise of rewilding is not just to return to a resilient nature that can recycle nutrients and energy efficiently without mining natural resources. This has to happen fast.

There are two critical reasons for this. The first is that rewilding will only work if there are multi-use systems. Some of the biomass still has to feed humans, and it will be hard to do this whilst rewilding is happening, but it will be a little easier when nature is back in balance. The second reason is that we want rewilding to help with other environmental issues, especially climate and water management. For example, to prevent some of the spiralling sequences that will accelerate global warming, carbon dioxide in the atmosphere must be sucked down as fast as possible into vegetation and soil. In some habitats, the level of change humans have wrought is such that the habitats are barren. Any recovery that nature might deliver will be slow. Rewilding in such cases must be a very active process and may not be possible. The technical details of how to speed up and give nature a lift are complex. Even knowing where we need to intervene because nature would otherwise take a long time to reverse the damage to the soil and vegetation. Instead of nature doing its thing, the need for speed means active management, which is the opposite of the intent. As we already know, being active comes with consequences.

Active management introduces cost and choice. What exactly should the active intervention be? Planting, weeding, fertiliser use, irrigation, flood control, animal reintroduction, fire management, and a host of tactic choices are around these broad strategies. Is there experience and expertise enough to make the right choice and implement them? And who pays for this activity? Whatever the objective, rewilding will cost money and require experts to help make management choices and skilled practitioners to deliver active management. In the practice of rewilding, there will be capability and capacity shortages.

Farmers have critical local knowledge of how to manage nature.

Regenerative agriculture

There are two broad modes of food production. Small-scale subsistence farming is practised by most farmers around the world, where the farmer produces enough food for himself and his family and little excess. The other production method is from a small number of farmers who produce intensively over large areas. They grow food for profit with significant inputs of energy and nutrients. These farms provide most of the food eaten by people worldwide who are not farmers, including the three and a half billion or so people living in towns and cities.

We can rewild both production systems by converging somewhere in the middle through various combinations of **regenerative agriculture**.

The principle is to have multiple crops maintained in closed-loop nutrient systems, in other words, to keep the nutrients within the landholding as much as possible. And have patterns of production that use plants and animals to operate more like an ecological system and less like an input-output system.

Conversion of most agricultural land to these practices will significantly influence landscape properties. It will improve the attributes of landscapes that will help the rewilding process. It will make agricultural land attractive to many organisms, much more so than currently. Critically, it will maintain soil biodiversity and enhance biological activity in the soil. And whilst it might not be truly wild in the way that Sir David would like it to be, it does provide habitat for a wide range of wild animals and plants. In other words, we have to change our approaches to agriculture, the subsistence approach and the intensification approach, to achieve rewilding effectively.

Is rewilding the solution? There is no doubt that biodiversity loss will continue at mass extinction rates unless something serious is done. Reserves are not cutting it, nor will any species-level initiatives. Wild animals and plants will need much more space to survive than what they currently have available or accessible.

Rewilding is a bold idea that can potentially integrate and return natural and agricultural landscapes closer to nature and reduce extinction rates. It can be the answer in some areas, especially those that have been degraded by human agricultural activity but are not suitable for food production in the long term or those areas that need rest from production. Abandoned agricultural land is a great place to start.

It may feel unattainable, especially in the form Sir David Attenborough suggests as a rewilding of one-third of the earth's land area. But the reality is that humanity has met equally unimaginable challenges in the past, and this one we have to achieve for any sustainable future. The big question is, if a third of the planet is rewilded, even with multiple use options, can the whole system deliver nutritious food equivalent to 22 trillion kilocalories daily for 100 years?

The value of audacity

In 1961, when the world struggled with east-west brinkmanship in the Cold War, U.S. President John F. Kennedy proposed an audacious national goal '*before this decade is out, of landing a man on the Moon and returning him safely to the Earth.*' Audacious barely describes such a call. The USSR had only just put the first man in space, computers were in their infancy, communications were limited, and the moon was far away. Imagine the details needed to get that job done. More than the technical difficulty, it was a bold shift in the rhetoric from nuclear armageddon to human achievement.

There are considerable challenges to returning a third of the planet to nature and feeding everyone. It is much more challenging than putting a man on the moon and returning him safely. But think how hard that was. The guidance computer on board Apollo 11 was puny—your mobile phone has millions of times more memory and processing power. Rewilding is even more audacious, and maybe that is its strength.

it compels public debate and collaborative discussions leading to revised agency policies to guide consistent and appropriate management direction.

Landres, P., Hahn, B. A., Biber, E., & Spencer, D. T. (2020). Protected area stewardship in the Anthropocene: Integrating science, law, and ethics to evaluate proposals for ecological restoration in wilderness. Restoration Ecology, 28(2), 315-327.

Rewilding is emotional and requires new paradigms. The practicalities are complex, and often conflicting values have to mesh. It sounds extreme, but rewilding 30% of the planet might be the answer to human sustainability.

Key Points

- Rewilding involves returning landscapes to a more natural state by removing human influences and reintroducing megafauna to restore ecosystem processes.
- Sir David Attenborough is a proponent of rewilding to solve the biodiversity crisis.
- Rewilding requires an ecological approach that considers energy, nutrient, and carbon cycles and balances wildlife conservation and sustainable food production.
- Rewilding has several challenges, such as deciding which areas to rewild, maintaining food production, addressing resistance from profit-driven landowners and the practicalities of imple-

mentation.

- Significant portions of the planet are already wild or unsuitable for agriculture, which should be considered in rewilding plans; whilst rewilding is a bold and logical concept, for success, it requires careful consideration, integration with food production, and a deep understanding of geopolitics.

In the next chapter

It is time we returned to saving the koala now that we know what extinction means and have an inkling of the fundamental issues and at least one of the solutions. It is time to look at dominion and making money.

Chapter Six

Saving the Koala

Wilderness may no longer exist because we have dominion. Manipulating nature into goods and services is what humans do well, better than any other organism. Can we keep doing it?

We love a landscape

I bet you know some of these names: John Constable, Thomas Cole, Joseph Turner, Claude Monet, and Vincent van Gogh. Yes, they are all famous painters and are especially remembered for their landscape paintings. You are also familiar with a painting by Constable, *The Haywain*, which depicts a rural scene on the River Stour between the English counties of Suffolk and Essex. Perhaps this one rings a distant bell, *A View of Arles* by Vincent van Gogh.

Many of the most famous landscape paintings are of rural scenes where human intervention has altered the land dramatically from the original '*wilderness*'. This was handy for the painters, of course. It meant they could see a long way into the distance as they stood at their easel or sketched their charcoal renderings. Their compositions could span across open fields dotted with human-made interests. Presumably, for the landscape painter, a pristine forest is less attractive visually, has fewer vantage points for the sketch, and, critically, has no apparent human connection. There are no objects or patterns to link the viewer to themselves.

I have already mentioned the biblical instruction in *Genesis 1:28* for humankind to have dominion, and over the centuries, humanity has readily complied. Almost all landscapes are altered by our hands, axes, and, more recently, chainsaws and D-9 bulldozers. What most of the famous landscape artists painted was a human-made landscape.

So why do landscapes excite us?

Landscapes are familiar to us. They represent the dominion we feel we have to have and reassure us with an image of perceived security that comes from power over nature; everything under our control makes us feel safer. If we control nature, then all her diabolical beasts, storms, and winter chills are less of a threat to our person. If we hold the profusion of life and can use it as needed, then we feel that security from food and water is easier to find. More viscerally, open land makes it easier to see the danger coming, the lion or the snake in the grass.

This rainforest image taken in the highlands of Papua New Guinea looks natural and pristine, but it has been altered by people planting trees they need closer to where they need them. These trees occur naturally in the forest, but their locations and abundance have changed over the thousands of years that people have been living in the forest. Even where we think it's natural, humans have their hand.

Dominion over nature

> *replenish the earth, and subdue it: and have dominion over the fish of the sea, and over the fowl of the air, and over every living thing that moveth upon the earth.*
>
> Genesis 1:28

It's an interesting injunction. It explains a lot about human behaviour, especially those in the West, many of whom default to the word of the Lord even though they have not been to church for a generation. The mandate sets up a war on nature because if we don't fight the wilderness, it will consume us; the evil forces of entropy will turn us into dust. We have to bring nature under our control by capturing its production for our food and fibre and removing the nasty predators and pests from our presence. We had a different approach to nature before agriculture. Aboriginal peoples did things very differently, even as they changed nature for their benefit. More than once, I have stood in a rainforest or an open savanna that modern science considers pristine. But the first peoples who lived there knew it was moulded by their ancestors who planted what they wanted and discouraged species with no apparent use or employed fire to alter the proportion of woody species. The result of millennia of human interventions is that the proportion of 'useful' plants is far more significant than by chance, not because people found new uses in all the different trees, shrubs and herbs, but because they increased the percentage of useful ones.

The same applies to the newly westernised landscapes of Australia. Before the Europeans came with metal tools, there is little doubt that the use of fire by aboriginal peoples gave the continent its eucalyptus forests dominated by tree species able to regenerate from seed or epicormic growth even after severe fire events.

Manipulation of nature is older than the plough and much older than the bulldozer and the tractor. However, we could borrow from these older, more benign ways of moulding nature that are more collective than individual. A forest dweller who sets a tree seedling to create a forest '*garden*' close to his village will not see the fruits of that tree. Even his son will need to wait until he is an elder. Collective use was intergenerational care, not something for today's market. But it suggests that it is possible to restore nature, to repair some of the damage done by capitalism. We certainly could do this restoration and rehabilitation of nature. Only there is a snag. We can't currently replace capitalism.

If we can understand the natural dynamics of the system and match the use of resources to that dynamic, then it is possible to return toward a wilderness. For example, we can rest fields and paddocks or lower expectations on production during a drought or allow regrowth rather than constantly '*tidying*' it up. These and many other actions could ease us away from dominion and closer to the alignment of production to the pulse of the natural world. However, replacing the '*imperative of individual enrichment*' is near impossible. This is as entrenched as any biblical edict. Nay, it even draws on it for sustenance. It would need a mighty, powerful political change to displace profit.

Food production for 8 billion human souls growing in number by over 70 million a year, and remembering that this means feeding them every day for at least 100 years will require intensive agriculture. There will be some dominion over the living things. And under our current paradigm, we will be doing it not with nature foremost but profit. We can make things better, and we must. We cannot do it from the premise that we fight nature.

We should also follow the biblical mandate and '*replenish the earth*'. We also cannot do this by assuming that we must return to wilderness nature, for that is not only a convenient myth but also a place that will not meet current demand. Ask any economist what happens to people when supply consistently fails to meet demand. It is never pretty.

It is prudent to concentrate on actions to 'replenish the earth', but until there is a viable alternative, we will have to use the forces of capitalism to that end. This is not an easy solution, for capitalism is corrupted by the '*individual enrichment*' drive that we all find hard to resist, but we need to figure out some options. Here is a small but effective one. Convert as much organic waste as possible into biochar by burning it without oxygen and then putting the biochar onto fields and pastures. This bumps carbon levels in the soil, increases water retention and aids nutrient exchange. Invest in the biochar plant and make money.

It sounds a bit whacky, but many left-field ideas like this sit alongside valid alternative agricultural paradigms (agroecology, regenerative agriculture, holistic management) and economics (degrowth, circular economies, doughnut economics). Alright, I have led you down a few garden paths that have little to do with saving the koala. I must show you that they are far more crucial to conserving this species than assertions of its imminent demise.

Why there is an environmental issue

Most items on a standard list of environmental issues are related to the loss or degradation of natural capital assets—pollution, soil erosion, pests, over-fishing, and waste. They occur when our actions and activities are not in balance with the ability of the natural world to absorb them.

Natural capital refers to the stock of renewable and non-renewable resources in the environment that contributes to the well-being of humans and the ecosystem. It encompasses many elements, including air, water, soil, biodiversity, and ecosystems. Essentially, natural capital represents the Earth's natural assets and the services it provides, such as clean air and water, pollination of crops, climate regulation, and habitat for various species. Recognizing the value of natural capital is crucial for sustainable development, as it involves understanding and managing these resources to ensure their preservation and responsible use, balancing human needs with environmental conservation.

When the yield of natural capital is compromised, the economic impacts can be acute, but the real problem is that these services are critical. So, the inescapable link between the environment and the economy is only the first layer of issues. The real problem is the **assumption that environmental assets are renewable** and capable of self-repair. When, in fact, we know that

- Critical thresholds exist when breached assets are depleted, sometimes precipitously—for example, fish stocks.

- Some assets are not renewable or renewable on timeframes, making them finite in management terms—minerals fall into this category, as does water in some areas.

- Once degraded, some assets cannot be restored or rehabilitated—once logged, a rainforest can only become a secondary forest, and extinct species are always extinct.

- Degraded assets, or those where the capital is used rather than the yield, impact future and current generations.

When natural capital is used, it tends to decline. It is hard to live off the interest. Typically, natural capital capability declines from its natural state when land is converted for agriculture. There is a change to the system's structure, and nutrients begin to be exported to the products. Natural capital capability declines to a base level under standard management practices. Where this level sits is a function of conditions and management practice, but the pattern is consistent.

Equally consistent is the ability to restore natural capital capacity with intelligent changes to management practices. Practices that retain ground cover promote soil structure, build moisture and encourage carbon retention can slow or reverse declines. Rebuilding natural capital is the best way to improve resilience and maximise future production potential.

Economic dependence on natural capital

Australians are fond of the phrase '*rode on the sheep's back*'. It evokes images of the rugged men and women who grew the wool that was the mainstay of the country's economy for generations. That outback toughness and resilience even symbolised what it was to be Australian.

Of course, the sheep rode the native vegetation of the land, helped along by fresh water pumped from aquifers of the great artesian basin. Today, mineral resources underpin the economy of Australia, but agriculture and agricultural products still account for around 12% of GDP or roughly $155 billion per year.

Many sectors of the economy rely directly on environmental assets for resources and raw materials such as water, timber and minerals that are required as inputs for producing goods and services—agriculture, forestry, food manufacturing, aquaculture, fishing, and tourism. Then, there are the sectors that support these production and use activities—finance, insurance, rural real estate, transport, and infrastructure. The whole economy is indirectly affected by ecosystem services, including carbon sequestration, water purification, managing flood risks, nutrient cycling and the many non-use values the environment offers.

Alright, but so what?

The retort from the conservationists to my tangent on natural capital will be something similar to the famous quote attributed to Albert Einstein: '*We cannot solve our problems with the same thinking we used when we created them*'. We commit the same sin if we make nature into goods and services and then apply the economic frame around them. Making money from nature is why the koala has a problem. We have to save the koala because capitalism is killing it.

Here, the mindful sceptic must take a deep breath. After being curious and assessing the evidence for the koala saga, we realised this was the surface issue, the emotionally charged fluff. The real challenge is resource use. First, the historical use that has converted so much of the planet's surface, depleted nature, and given us 8 billion humans, their livestock and their pets—the realities that are with us whether we like it or not. Second, there is a demand for natural capital from all these people.

People are the same everywhere. They care about themselves and their families a great deal, which takes up most of their capacity to care. They look after their pets and show some concern for their neighbours and members of their tribe, religion or ethnic group. Occasionally, they will help a stranger cross the street for no apparent reason. People care about cute animals, the Pope and the Denver Broncos.

Already a long way down the list with little caring left to give comes the planet, where everyone lives and provides all our food, water and shelter. Ask for help for the Earth, and people will offer it, so long as it does not compromise any of the items higher up on the carers list. This is the truth. And it just is. There is no moral judgment or banging of drums necessary. And it is not what people say. As we have seen, 71% of them tell the pollsters that they would prioritise environmental protection over job creation. They say the environment is important to them. And it is, along with all their other priorities.

A priority of caring has consequences. Any effort to sustain the planet's services must overcome this personal prejudice that people have, their NIMBY response, and the finite bandwidth for caring. Even if caring for the earth is right, it does not mean people will do it when they have other, hardwired priorities. We all should value the planet as a priority. If the services it provides fail, so do we. Everyone needs to know more, do more, make the sacrifices and stop making inane excuses. Rhetoric can be quite insidious. By definition, it does not need evidence or to be based on facts; it can be whatever the purveyor wants.

> *replenish the earth, and subdue it: and have dominion over the fish of the sea, and over the fowl of the air, and over every living thing that moveth upon the earth.*
>
> Genesis 1:28

Key Points

- Most famous landscape paintings depict scenes altered by human intervention, highlighting humanity's historical dominance over nature fuelled and justified by the biblical injunction from Genesis 1:28, which commands humanity to have '*dominion over the earth*'.

- The Western view of dominion over nature differs from that of Aboriginal peoples, who interacted with nature in more sustainable and collective ways but still altered the landscape.
- Capitalism's drive for individual enrichment challenges efforts to restore and rehabilitate nature, even though such restoration is essential.
- Conversion of land for agriculture typically leads to a decline in natural capital capability, altering the structure of ecosystems and exporting nutrients in products. Smart changes in land management practices can slow or reverse declines in natural capital, improving resilience and future production potential.
- There is tension between making money from nature and the imperative to conserve and protect the environment. Still, the core challenge lies in resource use—historical and ongoing conversion of natural landscapes, combined with the demand for natural capital driven by a growing global population.

In the next chapter

Saving the koala.

Chapter Seven

Making a donation

In the shopping malls and high streets of Australia, a passionate advocate often wears a bright-coloured t-shirt, a lanyard for authenticity, and a wide grin. She carries a white bucket only partially obscured by a sticker for whatever foundation or conservation organisation she represents for the day. The polite and usually young individual who is sometimes a volunteer and sometimes paid for her time will ask me for a donation to save the koala, just a few dollars.

"But I don't carry any dollars," I say, "I haven't for years."

"No problem, here, we have a mobile point of sale; you can just tap for a few dollars. It will save a koala."

In my mind, I have two questions: How and why?

How can a few dollars be spent to save a koala? What exactly is your organisation going to do with my money? Presumably, some of it pays for your time, your t-shirt, the many ads your folk run in the media, and the overheads to keep such a campaign alive. The rest goes to what? Planting trees, rescuing injured animals, putting up water points in the bush after wildfires... perhaps the preparation of erudite submissions to government committees. Knowing what happens to my donation when it leaves my account would be good.

Then, why should I pay? What evidence can you give me that your slogan 'Save the koala' is necessary? The koala is one of the hundreds of thousands of animals at risk of extinction; even humans fall into that category, for goodness sake, but somehow, the koala is unique. And anyway, aren't there koalas everywhere? I think I read there are populations from the Atherton Tableland west of Cairns in Queensland to islands off the coast of Victoria and South Australia in the south, and west to central and western Queensland, New South Wales and Victoria. The buggers are everywhere. Are you expecting all their habitat to be destroyed across a land area bigger than most countries? I need you to tell me why they need saving and from what?

In the real world, I said, "Thank you, but not today."

The look the youngster gives me from behind her polite smile is one of disdain that says you are one rich white fucker, so privileged you can't even offer a few dollars to save a koala; what a complete bastard you are.

I smile back and try not to let her see my anger.

She turns her smile to the next shopper, who eagerly taps their phone on the mobile point of sale.

Does she know what is going on?

The truth is that '*Saving the koala*' is a ruse, an obfuscation, a means to deflect attention from the actual problem. It is clever because it is emotionally charged, and that tap of the phone will send an endorphin shot to the brain of even the harshest sceptic. I would even feel good about it. Saving the koala or the panda or polar bear or any other creatures with fur used in these extinction statements is all because, to humans, they look cute, a bit like cuddly babies. We like them a lot. And when we want something or someone, it turns off many mental filters that protect us from scams and charlatans. The cuteness suckers us into many an irrational response.

This ruse is enough to turn me off all the green movements, for they all do it to some degree. They play on our emotional responses to further their agenda. But before I go on or push you away, I am not saying that the extinction of the koala or the polar bear would be desirable. The loss of biodiversity is a huge global challenge that must be addressed with all the will humanity can muster. And even though it's just one species, the koala represents biodiversity, the diversity from billions of years of evolution that humans are dismantling in a few generations.

When I stood beside the boma looking at that frightened rhinoceros in the Mana Pools National Park back in 1988, it occurred to me that I might be one of the last humans to see this species in a truly wild state. Separated from me by the sturdy mopane poles was an animal born in the wild and not in a game ranch or a protected reserve where progeny comes from a gene pool determined by human action and not the hand of god. Future rewilding notwithstanding, I might still be one of the few.

So, I get it. I understand the need for passion, urgency, and funds. I could stretch to let the end justify the means and forgive the obfuscation. I could, but I can't because saving the koala allows the passionate volunteer and the organisation she represents to ignore the critical problem and the primary cause of their angst: **wanton exploitation of nature to power the human enterprise.**

We are what we are, so we will always be saving the koala.

Science sources

All the *Mindful Sceptic Guides* are compiled from decades of experience with the science literature. Still, I also gathered specific *Evidence Reviews* that you can access on www.mindfulsceptics.info by clicking on or scanning the image below.

Mindful Sceptics

evidence reviews

scan or click the QR code for a deeper dive into the science sources for the Mindful Sceptic Guides

SCAN ME

About the Authors

Mark is an ecologist, not a green one. He has built, run and sold companies, held academic posts at four universities, advised governments on environmental matters, has a PhD and still plays too much golf. His ecological research ended up in more than 70 peer-reviewed research papers and over 80 technical reports, and he supplied peer-review for more than a dozen international science journals. He lived in Africa for a decade, where he researched termites and millipedes, and once walked into a lion who snarled at him. Much of his career has been in Australia, where he won an Outstanding Teacher award from Macquarie University.

His consultancy practice advises commercial and public sector clients on assessing, monitoring, and evaluating environmental impacts, especially in carbon asset and liability management. His non-fiction books are about awareness of humans' role in nature and, as Paul Sorol, he chronicles and writes fiction for a change.

Link in bio

Chris is an ecologist, a restoration one. A life-long mindful sceptic, Chris has happily landed as an ecological practitioner in his bush regeneration business. When not passionately managing land, being awarded Landcarer of the Year, trawling the evidence on nutrition, diet and health or carefully advising NGOs and government, he grows plants in his commercial nursery.

MINDFUL SCEPTIC GUIDES

Abatement

Biodiversity loss

Escaping the Malthusian trap

Farmers

Food security

Food, people and soil

Growing enough food

Healthy and sustainable diet

Sustainability

The soil carbon debt

mindfulsceptics.info

Made in the USA
Monee, IL
04 November 2026

67379715R00059